CONTENTS

D0187802

ALSO BY ALEX BELLOS

Can You Solve My Problems?

Puzzle Ninja

Here's Looking at Euclid

The Grapes of Math

Futebol: The Brazilian Way of Life

BY ALEX BELLOS AND EDMUND HARRISS

Patterns of the Universe:
A Coloring Adventure in Math and Beauty

Visions of the Universe:
A Coloring Journey through Math's
Great Mysteries

PERILOUS PROBLEMS FOR PUZZLE LOVERS

Math, Logic & Word Puzzles to Challenge Your Brain

ALEX BELLOS

THE EXPERIMENT

NEW YORK

To Natalie

PERILOUS PROBLEMS FOR PUZZLE LOVERS: *Math, Logic & Word Puzzles to Challenge Your Brain*
Copyright © 2019, 2020 by Alex Bellos

Originally published in Great Britain as *So You Think You've Got Problems?* by Guardian Books/Faber and Faber Ltd in 2019. First published in North America in revised form by The Experiment, LLC, in 2020.

The Experiment, LLC
220 East 23rd Street, Suite 600
New York, NY 10010-4658
theexperimentpublishing.com

THE EXPERIMENT and its colophon are registered trademarks of The Experiment, LLC. Many of the designations used by manufacturers and sellers to distinguish their products are claimed as trademarks. Where those designations appear in this book and The Experiment was aware of a trademark claim, the designations have been capitalized.

The Experiment's books are available at special discounts when purchased in bulk for premiums and sales promotions as well as for fund-raising or educational use. For details, contact us at info@theexperimentpublishing.com.

Library of Congress Cataloging-in-Publication Data

Names: Bellos, Alex, 1969- author.
Title: Perilous problems for puzzle lovers : math, logic, & word puzzles to
   challenge your brain / Alex Bellos.
Description: New York : The Experiment, 2020. | Originally published in
   Great Britain as So You Think You've Got Problems? by Guardian
   Books/Faber and Faber Ltd in 2019.
Identifiers: LCCN 2020037429 (print) | LCCN 2020037430 (ebook) | ISBN
   9781615197187 (paperback) | ISBN 9781615197194 (ebook)
Subjects: LCSH: Mathematical recreations. | Puzzles.
Classification: LCC QA95 .B4416 2020 (print) | LCC QA95 (ebook) | DDC
   793.74--dc23
LC record available at https://lccn.loc.gov/2020037429
LC ebook record available at https://lccn.loc.gov/2020037430

ISBN 978-1-61519-718-7
Ebook ISBN 978-1-61519-719-4

Cover and additional text design by Jack Dunnington
Text design by carrdesignstudio.com
Illustrations by Andri Johannsson
Cartoons by Simon Landrein
Images for problem 99 copyright © Alain Nicolas
Back cover "folded letter" puzzle copyright © Scott Kim

Manufactured in the United States of America

First printing October 2020
10 9 8 7 6 5 4 3 2 1

INTRODUCTION

Archimedes was the greatest scientist of antiquity. He made stunning theoretical discoveries about concepts such as pi and infinity, and designed some of the most advanced technology of his age.

He is also responsible for one of the worst puzzles in history.

It's a stinker, believe me.

The "cattle problem" is difficult, inelegant, and absurd. It is nevertheless a perfect place to begin a book of recreational problems. For a start: Archimedes. Secondly, it is a fascinating historical oddity. Lastly, bad puzzles illuminate what makes good ones good. Our dissection of the Archimedean herd explains what the rest of the puzzles in this book will *not* be like. You will be grateful.

THE CATTLE PROBLEM

The sun god had a herd of cattle that grazed on the plains of Sicily. The bulls and cows came in four colors: white, black, yellow, and dappled, such that the numbers of each type could be expressed in the following way.

White bulls = ($\frac{1}{2} + \frac{1}{3}$) black bulls + yellow bulls,

Black bulls = ($\frac{1}{4} + \frac{1}{5}$) dappled bulls + yellow bulls,

Dappled bulls = ($\frac{1}{6} + \frac{1}{7}$) white bulls + yellow bulls,

White cows = ($\frac{1}{3} + \frac{1}{4}$) black herd,

Black cows = ($\frac{1}{4} + \frac{1}{5}$) dappled herd,

Dappled cows = ($\frac{1}{5} + \frac{1}{6}$) yellow herd,

Yellow cows = ($\frac{1}{6} + \frac{1}{7}$) white herd,

White bulls + black bulls = a square number,

Dappled bulls + yellow bulls = a triangular number.

What is the total size of the herd?

Urgh. Before we try to digest this unappetizing soup of fractions, here's its curious backstory: The problem was discovered in a library in Germany in the eighteenth century, 2,000 years after Archimedes died. Written in Greek in the form of a poem made up of 22 couplets, it was found in a manuscript that no one had looked at before, accompanied by a note stating that Archimedes had sent it to Eratosthenes, head of the library at Alexandria.

On the positive side, the problem is set in verse. At least its author aimed to entertain. The math, on the other hand, is less jolly. The cattle problem demands a farmyard of algebra. The first seven lines can be written as seven equations in eight unknowns. With enough patience and enough paper, after much tedious computation and shuffling around of variables, you will discover that the smallest possible size of the herd that satisfies the first seven lines is 50,389,082. (Which means Sicily would have a bull or a cow for every 500 square meters.)

If you've solved the puzzle thus far, Archimedes congratulates you. But don't get smug. "Thou . . . can not be regarded as of high skill," he warns. We've not got to the hard bit yet.

The eighth line states that the number of white and black bulls is a square number, meaning a number like 1, 4, 9, or 16 that is the square of another number (i.e., 1^2, 2^2, 3^2, 4^2). If this property is included, the smallest possible size of the herd is 51,285,802,909,803. (Sicily now has about 2,000 animals per square meter, meaning that the island is entirely covered with cattle, squashed like sardines and stacked hundreds of meters high.) This calculation requires a little bit more advanced algebra, but not too much. Gotthold Ephraim Lessing, the German librarian who discovered the problem, showed it to a mathematician friend who was able to come up with this solution.

The final line is the killer. It states that the number of dappled and yellow bulls is a triangular number, meaning a number that can be arranged in a dot triangle, such as 3, 6, 10—as in ●●, ●●●, ●●●●—and so on, with an extra line each time. Bam! Archimedes's cattle problem is now beyond the scope of eighteenth-century mathematics.

For the next hundred years, the bovine brainteaser was a celebrated unsolved problem. It was rumored that Carl Friedrich Gauss, the greatest mathematician of the nineteenth century, had solved it completely. The first person to publish a partial solution, however, was fellow German August Amthor in 1880, who revealed that the smallest possible herd was a number beginning with 766 and continuing for another 206,542 digits. In other words, a number so ridiculously huge that the universe would not be able to contain this herd even if every bull and cow was as small as an atom.

Undeterred by the scale of the job, in 1889 three friends in Illinois with nothing better to do started to work out the other digits. After four years' toil they had calculated 32 of the digits on the left side of the number and 12 on the right. The full solution to the cattle problem, however, required the arrival of the computer age. In 1965 a supercomputer took 7 hours and 45 minutes to print out the number, which ran to 42 sheets of A4 paper.

Lessing and others have questioned whether Archimedes was indeed the author of the cattle problem. No reference to the puzzle exists in any other Greek writing, and Archimedes could not possibly have known the answer to the question he supposedly set. Yet some academics are convinced it does date to him. Archimedes was fascinated by extraordinarily large numbers; in his short text *The Sand Reckoner*, he devises a new number system in order to estimate the number of grains of sand that would fill the universe. (His estimate: 10^{63} grains.) Perhaps the point of the cattle problem was not to solve it at all, but to show how nine simple statements using unit fractions could determine a number that (in Archimedes's time) was beyond all comprehension. To concoct a whimsical, easily understandable problem that nevertheless remains unsolved for more than 2,000 years is arguably the mark of (evil) genius. *If thou hast computed [the answer], O friend, and found the total number of cattle,* ends the verse, *then exult as a conqueror, for thou hast proved thyself most skilled in numbers.* Quite.

As a puzzle, the cattle problem is less a piece of recreational math than an overly complicated exercise in solving simultaneous equations.

The remaining puzzles in this book:

Prize insight over computation.

Engage basic competence rather than technical skills.

Deal in numbers that can be written down on fewer than 42 sheets of A4 paper.

Can be fully solved in less than 2,000 years.

I'll take my lead from Archimedes in only one way: his zoological choice of subject matter.

This book kicks off with a chapter of puzzles about animals. Randy rabbits, mischievous tabbies, frogs, flies, lions, camels, chameleons, and more. Animal puzzles do not—yet—constitute a mathematical field of their own, but they do provide a delightful and diverse bestiary of brainteasers, allowing me to showcase some of my favorite puzzles from the Middle Ages to the present day.

After visiting the animal kingdom, we will find ourselves in peril. In real life, you may never have been abandoned on an island, trapped in a maze, locked in a room, or stuck on death row. In puzzle-land, however, we get into these scrapes *all the time*, as you will discover in the second chapter, in which the problems are concerned with escape and survival. You will be required to think laterally, logically, and even topologically. Several of these puzzles are based on fascinating discoveries in computer science, in which devising a strategy to get out of jail, say, is analogous to building an algorithm.

I've written this book to share the joy I get from solving problems. A good puzzle will not only stimulate creative thinking, but also spark a sense of wonder and curiosity about the world. I have taken care to choose questions that present the solver with a surprise, or which reveal an interesting pattern or idea. Puzzles are a versatile medium, covering a huge variety of genres, and I hope this book will tickle your brain from all sides.

The puzzles are not organized by level of difficulty. You can read the chapters from beginning to end or skim them and pick and choose. I include material about the history of mathematics, and the role of puzzles within it, and have full explanations and follow-up discussions in the back. The problems in the "Tasty teasers" sections are snacks to get you in the mood.

Indeed, you will have nibbled a puzzle already, the one on the back cover about the folded shape. I love this puzzle because the first letter that comes to mind when you try to unfold the shape is an L, which is the wrong answer. It takes some mental effort to discard the obvious letter, at which point a flash of insight may reveal the less obvious one. Puzzles often play with our minds this way, deliberately leading us up the garden path, or presenting us with a tantalizing, but entirely erroneous, solution. The pleasure in solving a puzzle that is trying to misdirect makes the final "aha!" especially sweet.

The area of mathematics in which we are most handicapped by our own psychology is probability, the theme of the final chapter. Our brains are poorly equipped to understand randomness, and probability puzzles are a great way to identify where our intuitions go wrong. Not only do these puzzles surprise and enlighten us, they also help us think more clearly.

Indeed, that is the power of all puzzles. They are fun. But they are also useful. Puzzles make our brains more nimble, versatile, flexible, and multifaceted. They improve our capacity for reasoning, hone our ability to spot patterns, train us to look at the world from different perspectives, and point out areas where we are easily misled.

Now pack your bags.

The animals are eager to meet you.

Tasty teasers

Number conundrums

1)

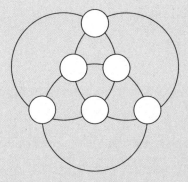

Each of the three big circles passes through four small white circles. Place the digits from 1 to 6 in the white circles so that the numbers on each big circle add up to 14.

2)

Place the digits from 1 to 8 in the circles such that no number is connected by a line to a number either 1 more or 1 less than itself. For instance, 6 cannot be connected to 5 or 7.

3)

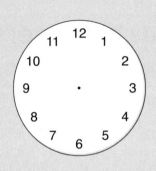

Divide this clock face with two straight lines so that the sum of the numbers in each section is the same.

4) Place the digits indicated in the squares so that each equation makes sense. For example, the first one uses the digits 1, 2, 3, and 4.

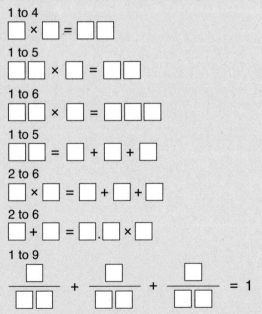

1 to 4

☐ × ☐ = ☐☐

1 to 5

☐☐ × ☐ = ☐☐

1 to 6

☐☐ × ☐ = ☐☐☐

1 to 5

☐☐ = ☐ + ☐ + ☐

2 to 6

☐ × ☐ = ☐ + ☐ + ☐

2 to 6

☐ + ☐ = ☐.☐ × ☐

1 to 9

$$\frac{☐}{☐☐} + \frac{☐}{☐☐} + \frac{☐}{☐☐} = 1$$

5)

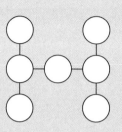

Place seven digits from 1 to 9 in the circles so that the three digits on each line multiply to the same amount.

6)

In this triangle, the digits from 1 to 6 are positioned such that the difference between two adjacent numbers is shown in the row beneath them.

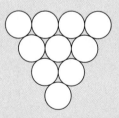

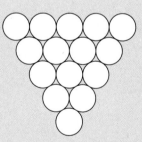

Do the same for the digits 1 to 10 in the triangle on the left, and the digits 1 to 15 in the triangle on the right.

The puzzle zoo

ANIMAL PROBLEMS

THE THREE RABBITS

Can you rearrange the positions of the three rabbits and the three ears so that each rabbit has two ears?

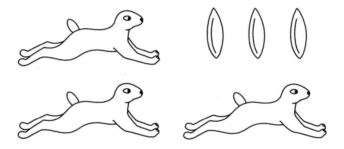

If you're reading this in a Devon church, don't look up!

More than a dozen churches in Devon, England, have medieval wooden carvings on their ceilings that display the solution. In fact, the image of three rabbits, or hares, sharing their ears is a symbol that appears in many sacred sites across the northern hemisphere, the earliest dating from sixth-century China. The largest cluster of examples, however, is in Devon, where the linked leporine lugs are sometimes known as the Tinners' Rabbits, possibly because wealth from local tin mines helped build and maintain the churches hundreds of years ago.

The simple symmetry of the three rabbits makes them a powerful mystical symbol, an easy metaphor for ideas of eternity and beauty. Yet part of their allure is their elegance as a puzzle, in which the image makes sense when seen one way, but not when seen in another. We are attracted to images that play with our sense of perception. Puzzles are by their nature mesmerizing; they spin our heads right round.

DEAD OR ALIVE

These dogs are dead you well may say:
Add four lines more, they'll run away!

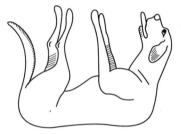

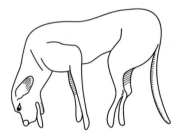

This canine conundrum dates from 1849, when it appeared in the first issue of *The Family Friend*, a lifestyle magazine in book form aimed at the Victorian housewife. It's the original version of a visual illusion that has been reinvented many times since, in which you must draw four lines on an image in order to bring two animals back to life.

Speaking of rabbits and dogs . . .

GOOD NEIGHBORS

A woman opened the door and her dog walked in. The dog had the neighbor's pet rabbit in its mouth. The rabbit was dead. Distraught, the woman went next door immediately to apologize. The neighbor smiled. "Don't worry, my rabbit was not harmed."

Why was the rabbit not harmed?

The mysticism around the three rabbits also turns on rabbit behavior in the wild. The breeding rate of bunnies has made them age-old symbols of fertility and rebirth, anthropomorphic shorthand for the highly sexed. Prolific procreation is their principal protection against a profusion of prowling predators. Reproduction is ripe for mathematical analysis. Indeed, one of the most famous problems in mathematical literature is about a rapidly expanding rabbit family.

In the *Liber Abaci*, the thirteenth-century book that introduced Arabic numerals to Europe, Leonardo of Pisa, better known as Fibonacci, set the following problem. Start with a pair of rabbits. If they produce a pair of offspring every month, and if every new pair becomes fertile after a month, again producing another pair every month, how many pairs are there at the end of 12 months? Saving you the bother of calculation, the monthly totals are (1), 2, 3, 5, 8, 13, 21, 34, 55, 89, 144, 233, and 377, an ordered list of numbers known—thanks to this problem—as the Fibonacci sequence.

(We start with 1 pair. At the end of the first month, this pair has bred, so we have 2 pairs. At the end of the second month, the first pair has bred again but the second pair is not yet fertile, so we have a total of 3 pairs. At the end of the third month, the first and second pairs have now bred, but the third is not yet fertile, giving a total of 5 pairs. And so on.)

The Fibonacci sequence is one of the few number sequences known beyond mathematics. It has many intriguing properties. For example, each term is the sum of the previous two ($1 + 2 = 3$, $2 + 3 = 5$, $3 + 5 = 8$, and so on). This recursive process models natural growth—for plants as well as buck-toothed mammals—which is why the number of spirals in cauliflower, Romanesco broccoli, pine cones, pineapples, and sunflower heads is nearly always a number in the Fibonacci sequence. Do check, next time you're at the supermarket.

Fibonacci hit on a fascinating piece of pure mathematics. Yet how well does his problem reflect the real world? In other words, did he create an accurate model of the reproductive practices of rabbits?

To Fibonacci's credit, he got the gestation period about right. Rabbits take a month to have babies, and can be impregnated again within minutes of giving birth. Poor things. So, theoretically, a doe can have 12 litters a year. On the other hand, she becomes fertile more slowly, after about 6 months, and she has more children, on average about 6 kits per litter. Armed with the relevant zoological data we can update Fibonacci's historic rabbit riddle.

A FERTILE FAMILY

How many descendants does a female rabbit have in her lifetime if:
Rabbits become fertile after 6 months.
Once fertile, a doe produces every month a litter of 6 kits, 3 of which are female.
The life span of a rabbit is 7 years.

Of course, these details still paint a simplified picture. In the wild, the life span of a rabbit is only about a year. After a few years the fertility of female rabbits drops off. Environmental factors such as available space and food will limit the rate of growth. Nevertheless, the question provides a theoretical estimate of the upper limit of potential bunny reproduction, a scientific analysis of what it really means to breed like rabbits.

I'll give you full points if you can work out the formula for how to calculate the answer. To get the exact figure you might need some computer assistance. For those who aren't masters of technology (or Excel), look in the back. But before you do, estimate what you think the answer is. If you get it right to within two powers of 10—i.e., up to 100 times more, or 100 times less—treat yourself to a meal of *lapin à la moutarde* and a bottle of Chablis. You will be amazed.

The next puzzle is about rabbits. Sorry, I mean *ribbits*.

(5)

A BUNCH OF HOPS

10 lily pads are positioned in a straight line across a pond.
A frog sits on the leftmost lily.

At any stage, the frog can either jump to the next lily along, or hop over that lily and land on the lily 2 positions along.

If the frog never goes backward, how many different ways are there for it to reach the lily on the right?

Let's hop on. From an animal famed for its jumps, to one famed for its humps.

The camel is a double celebrity in puzzle-land.

First, it gives a bravura performance as the protagonist in a medieval puzzle about transporting grain. Secondly, it appears in a traditional puzzle about a family feud. (I'll start with the first and get to the second later.)

The gist of the grain transportation problem is this: What is the best strategy to get a camel to deliver grain from A to B, given that the more the camel walks, the more of the grain it will require as food? The first question of this type appeared in the foundational work of modern-day puzzledom, Alcuin of York's eighth-century manuscript, *Propositiones ad Acuendos Juvenes*, or *Problems to Sharpen the Young*. A caravan of similar problems about walking and eating, or walking and drinking, trod over the dunes in the following centuries.

(6)

CROSSING THE DESERT

Four Bedouin tribesmen, each with a camel, are standing together at the edge of a desert. The group must deliver an important package to a camp in the middle of the desert, 4 days' camel-ride away. Camels can only carry enough water for 5 days. If the Bedouin cooperate, and are able to transfer water between camels in the desert without losing any due to spillage and evaporation, how is it possible for one of them to deliver the package and all of them to return to their starting point with only 20 days' supply of water?

Alcuin's camel problem is a brilliant example of a piece of fun that evolved into a serious area of mathematical research. In the twentieth century, the grain-guzzling camel was upgraded to a gas-guzzling machine. The "jeep problem" asks you to find the best way to drive as far as you can from your source of gasoline if you can only carry a finite amount of gasoline at a time, but are allowed to deposit it at drop-off points and return to pick up more. This problem has obvious applications in exploration and warfare. Indeed, the first detailed analysis of the problem was funded in 1946 by the United States Army Air Forces. If your exploratory mission requires you to bring your own fuel—because you are traversing the Antarctic, flying over enemy territory, or discovering new areas of the solar system—you will grapple with exactly this type of logistical issue.

⑦

SAVE THE ANTELOPE

You are a vet working in the Sahara when you hear that an endangered antelope has broken its leg 400 miles away from your practice. You decide to rescue the animal in your jeep, the specifications of which are:

   It does 100 miles to the gallon.

   The tank has a maximum capacity of a gallon of gasoline.

   In addition to what's in the tank, the jeep will carry four one-gallon canisters of gasoline.

   There are no gasoline stations between you and the injured animal, so in order to drive long distances you have to drop off canisters at certain points and return to pick them up later. You can only drop off canisters that are full. You can return to base as many times as you like to refuel.

   How do you reach the animal and bring it to your practice using only 14 gallons of gasoline?

Study of the jeep problem led to a staggering result, the kind you can't quite believe even though the math proves it's true. Let's imagine there is a service station with an unlimited supply of fuel, and you have a jeep with a finitely sized tank. It is possible to drive *as far as you like* from that service station using only gasoline from there. In other words, you could theoretically drive all the way around the world in a Ford Focus using only gasoline that you took with you from New York. Like the problem above, the strategy relies on making many trips to drop off fuel and pick it up later. If n is the number of miles you can drive on a full tank, you can get as far as n with no drop-offs. You'll have to take my word that you can reach a distance of $(1 + \frac{1}{3})n$ with a single fuel drop-off, you can reach $(1 + \frac{1}{3} + \frac{1}{5})n$ with two drop-offs, and the more drop-offs you are able to make, the farther and farther you can travel, even though the extra distances you can go get smaller and smaller. Since the series $1 + \frac{1}{3} + \frac{1}{5} + \frac{1}{7} + \ldots$ is a divergent series, meaning that by including more and more terms it can be made to exceed any finite value, the distance the jeep can travel will also exceed any finite value.

The *other* classic problem about camels involves three children arguing about their inheritance. The kids are in a hump about who gets the humps. The puzzle in the form presented below dates from the nineteenth century. In recent years it has been reinvented as a morality tale about how a random act of generosity solves an apparently intractable problem.

A man's dying wish is that his herd of 17 camels be divided among his three children, such that the eldest child gets half of them, the middle child gets a third of them, and the youngest a ninth of them. The children are unable to decide how many camels each is to receive because of the arithmetical impossibility of dividing 17 by 2, 3, or 9 and getting a whole number of camels. (No camels are to be harmed during the solving of this problem.)

In order to resolve the dispute, the children approach a wise old woman and explain the situation. She listens intently. To the children's surprise, she fetches her own camel and gives it to them.

"Now that you have 18 camels," she says, "you can divide them up in accordance with your father's wishes."

The eldest child takes 9 camels, which is half of them, the middle child takes 6, which is a third, and the youngest takes 2, a ninth. However, 9 camels + 6 camels + 2 camels = 17 camels. In other words, one camel is left over. "I'll take my camel back, thanks," says the old woman, and she and her camel walk away.

The puzzle here is to explain the apparent contradiction in why a group of objects that cannot be cleanly divided into a half, a third, and a ninth can be divided into those proportions when an extra object is added and then taken away. (I'll explain how it works in the Answers section.)

The wise woman who negotiates peace among feuding siblings makes for a charming parable, in which the eighteenth camel represents a fresh idea that resolves a deadlocked situation.

She returns to help another family in a similar predicament.

(8)

THE THIRTEEN CAMELS

A man leaves 13 camels to his three children in the following proportions: to the eldest, half; to the middle child, a third; to the youngest, a quarter. The children can't decide how many each should get, because you cannot divide 13 by 2, 3, or 4 without inflicting severe pain on a camel.

They consult a wise old woman to resolve their dispute. How does she fix it?

Camels and horses are historically the two most popular riding animals. Have you ever wondered which is faster? Or slower?

⑨

CAMEL VS. HORSE

Kamal has a camel and Horace has a horse. The friends are arguing about which animal is slower, so they decide to race them along a one-mile stretch of track, with the winner being the one that reaches the finish line last. They get in their saddles. But, predictably, they just stand there, since no one wants to start first and risk being the first to finish. An hour later, Ada shows up. She asks what's going on. The men get out of their saddles and explain. Ada says a few words, at which point the men sprint to the animals, jump on, and race to the finish line as fast as possible.

What was Ada's advice?

The following animal puzzle also involves two friends riding at speed in a straight line.

⑩

THE ZIG-ZAGGING FLY

Two cyclists are racing toward each other along a straight road. When they are 20 miles apart, a fly on the nose of one of the cyclists starts flying in a straight line toward the nose of the other cyclist. When it reaches the nose of the other cyclist, it immediately turns around and flies back toward the first cyclist, and it carries on flying between the cyclists' noses as they approach each other.

If the cyclists are both cycling at a constant speed of 10 miles an hour, and the fly flies at a constant speed of 15 miles an hour, how far has the fly flown when the cyclists meet?

You can solve this the hard way or the easy way. The hard way is to calculate how far the fly travels before it touches the nose of the second cyclist, and then how far it travels back before touching the nose of the first cyclist, and so on, adding up a series of decreasing lengths.

I'll leave it to you to find the easy way.

The zig-zagging fly is part of mathematical folklore because of an episode involving one of the twentieth century's greatest scientists, John von Neumann, the Hungarian-American who made many important advances in economics, computer science, and physics.

On being told the puzzle by a friend, von Neumann solved it instantly in his head.

"So you saw the trick," his friend remarked.

"No," he replied. "I just added up the distances."

Geniuses can sometimes be a bit stupid.

The next puzzle is also about insects moving in one dimension. And like the last one, its apparent unwieldiness unravels upon a simple insight.

(11)

THE ANTS ON A STICK

Six ants are walking along the edge of a 1-m stick, as illustrated below. Aggie, Bozo, Daz, and Ezra are walking from left to right as we look at the diagram. Carlos and Freya are walking from right to left. The ants always walk at exactly 1 cm per second. Whenever they bump into another ant, they immediately turn around and walk in the other direction. When they get to either end of the stick, they fall off.

Their starting positions from the left end of the stick are: Aggie 0 cm, Bozo 20 cm, Daz 38.5 cm, Ezra 65.4 cm, and Freya 90.8 cm. Carlos's position is not known—all we know is that he starts somewhere between Bozo and Daz.

Which ant is the last to fall off the stick? And how long will it be before he or she does fall off?

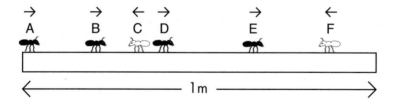

Now for a puzzle involving rubber. It's stretching, that's for sure. Like the previous question, it is also about invertebrates traveling from one end of a long piece of material to the other.

THE SNAIL ON THE RUBBER BAND

A snail lies at one end of an rubber band that is 1 km long, as shown below. It crawls toward the other end at a constant rate of 1 cm per second. At the end of each second, the rubber band is stretched by a kilometer. In other words, when the snail has crawled 1 cm the band is 2 km long, when it has crawled 2 cm the band is 3 km long, and so on.

Show how the snail eventually reaches the end of the rubber band.

The snail sets off on a daunting task that appears to get even more dispiriting as each second passes. If it travels at only 1 cm a second, and the rubber band stretches by 1 km a second, our initial reaction is that the snail will always be getting farther and farther from its target destination rather than nearer to it. It's a beautiful puzzle, however, because the snail does get there in the end. (We need to assume that the rubber band stretches uniformly as far as possible, and that the snail never dies.)

To get a sense of why the snail is not doomed to eternal crawling, consider its distance from the left end of the rubber band. (It's a mathematical snail, so consider it as a point, starting at the band's left edge.) After one second, the snail has traveled to a point 1 cm along, which on stretching becomes a point 2 cm along, since stretching the band from 1 km to 2 km has the effect of doubling the distance between any two points. After another second the snail is 3 cm from the left end, which on stretching instantly becomes 4.5 cm, since stretching the band from 2 km to 3 km has the effect of multiplying the distance between any two points by 3/2. In other words, the snail is carried forward by the stretching and covers an increasing distance each second, which gives us some hope that it might be able to traverse its ever-expanding floor.

I included the snail puzzle because the result is spectacular, even though the full proof requires a piece of information that, while familiar to mathematicians, will not be known to everyone. Astute readers will notice, however, that this result can be deduced from material I have already shared in preceding pages.

Now for some puzzles that require no technical mathematics at all.

(13)

ANIMALS THAT TURN HEADS

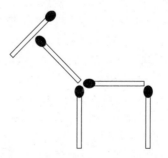

Move a single matchstick so that the horse changes direction.

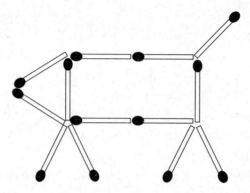

The dog is facing left. Can you make him face right—while keeping his tail pointing upward—by moving only two matches?

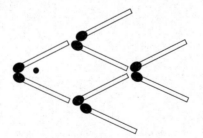

The fish has a blueberry for an eye. Can you move the blueberry and three matches to make the fish point the other way?

(14)

BANISHING BUGS FROM THE BED

Your bedroom is full of critters that can wriggle and creep along any solid surface, but cannot swim.

You want to stop these pesky bugs getting into your bed. It's easy to stop them getting up from the floor: You stand each leg of the bed in a bucket of water.

But how do you stop them from crawling up to the ceiling and falling down on your sheets? A gutter like the one shown below won't work since the bugs could fall onto the edge of the gutter and then crawl around and drop onto the bed.

What construction will keep your bed bug-free?

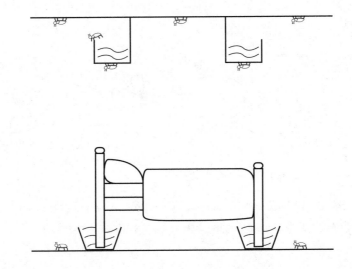

Suspending a canopy, or mosquito net, over the bed is not the solution, since the critters will crawl all over it, and might well crawl underneath it and up to the bed. Were you to find a way to hermetically seal this canopy to the floor, you would still have to open it to reach the bed, and when you did the bugs would be able to crawl through the opening too.

The next puzzle might sound like a sitcom skit, but it has genuine mathematical content. Everyday language is full of ambiguities and assumed knowledge. Mathematical statements, on the other hand, are precise. The challenge of the puzzle is to apply mathematical rigor to a non-mathematical statement for comic effect. Let your inner pedant run free!

(15)

THE DUMB PARROT

The owner of a pet shop never lies and is very precise with his words. A customer asks him about the parrot in a cage on his counter. "This bird is extremely intelligent," he replies. "She will repeat every word that she hears." The customer buys the bird. A few days later, however, the customer returns. "I'm furious! I spoke to the parrot for hours but the stupid bird has not repeated a single thing!"

Given that the pet shop owner did not lie, how is this possible?

Here's one for nothing. The parrot could be dead. The Monty Python solution. Easy. But can you come up with at least *four* more reasons for the bird's silence that—while possibly far-fetched—do not contradict the owner of the shop?

So far, I've included puzzles involving mammals, arthropods, a fish, an amphibian, and a bird. Only one class of animal is left.

(16)

CHAMELEON CAROUSEL

A colony of chameleons on an island currently comprises 13 green, 15 blue, and 17 red individuals. When two chameleons of different colors meet, they both change their colors to the third color. Is it possible that all chameleons in the colony eventually have the same color?

Heron of Alexandria lived in the first century BCE. He is the most important mathematician to share a name with a common animal, which is surely enough to get him a mention in this chapter. Heron invented many ingenious contraptions that were way ahead of their time, including a vending machine, a mechanical puppet theater, and a steam engine. He also discovered a theorem that is the basis of many splendid puzzles. I'll illustrate it here in the context of two houses and a road. It states that the shortest path from house A to house B via a point on the road is the one marked below, constructed by drawing a line to B', where B' is a reflection of B across the road. The line from A to B' is straight, so it is obviously the shortest path from A to B', and it is also the same length as the path from A to B via the road. You are now equipped to solve the next puzzle, about an animal taking the optimal path to lunch.

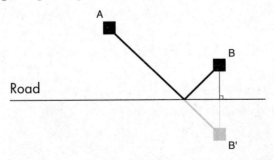

(17)

THE SPIDER AND THE FLY

A fly lands on the inside of a cylindrical glass tumbler, 2 cm from the rim, as illustrated right. A spider is on the opposite side, 2 cm from the bottom, on the outside of the glass. The glass has a height of 8 cm and a circumference of 12 cm. If the fly remains stationary, what is the shortest distance the spider must crawl in order to reach it?

Look again at the earlier image of the houses and the road. The angle at which the path from A hits the road is the same as the angle at which the path leaves the road for B. In other words, the picture illustrates the law of reflection of light, which states that when light hits a mirror, the angle of incidence is equal to the angle of reflection. (Imagine the road is a side view of a mirror, and A is a light source. The beam that leaves A and reflects to B is exactly the one illustrated by the black line.) Heron knew the law of reflection of light. Using his theorem about minimal distance, he became the first person to deduce that light always takes the shortest path.

(18)

THE MEERKAT IN THE MIRROR

A meerkat is looking at itself in a wall mirror. It sees its reflection perfectly framed: The top of its head reaches the top of the mirror and its feet reach the bottom.

What happens when the meerkat steps back from the mirror? Does it see less of itself in the mirror, or do gaps open up above its head and beneath its toes?

Assume that the wall mirror is vertical, and that the meerkat is rod-straight upright when looking at itself.

Meerkats lend themselves well to the previous problem since the one thing everyone knows about them is that they like to stand up. What about the behavioral traits of *real* cats? They are mischievous, independently minded, and like to move around at night.

(19)

CATCH THE CAT

A straight corridor has seven doors along one side. Behind one of the doors sits a cat. Your mission is to find the cat by opening the correct door. Each day you can open only one door. If the cat is there, you win. If the cat is not there, the door closes, and you must wait until the next day before you can open a door again. The cat is restless and every night it moves to sit behind another door. The door it moves to is either the one immediately to the left or the one immediately to the right of where it was previously.

How many days do you need to make sure you will find the cat?

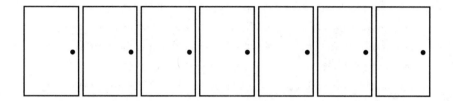

The question is asking you to find a strategy that guarantees you will catch the cat in a fixed number of days, whichever door it starts behind and wherever it moves in the night. The key to the solution is to begin with a smaller number of doors, find the pattern, and then increase the number of doors. I'll get you started. Imagine there are only three doors. If you open the middle door on two consecutive days you are guaranteed to get the cat, since if the cat is not behind the middle door on day one, it must be behind either of the end doors. And if it is behind either of the end doors on day one, it has no option but to move behind the middle door on day two. If there are four doors, you can catch the cat in four days. I'm not going to explain the strategy, but you will purr with delight when you work it out. Remember, the cat will only move to a door immediately to its right or its left, and it may return to a door it was behind previously.

Predatory aggression is a common theme in puzzles—usually, but not always, felt by frustrated solvers.

MAN SPITES DOG

A house is surrounded by a five-foot-high wall. Its front door can only be reached by a path, which you enter from the main gate.

A mailman arrives at the gate. He sees a dog in the garden. The dog sees him, springs up and runs to the gate to attack him, stopped only by its leash, which is tied to a tree. The dog is barking and straining at the leash, trying to get as near to the mailman as possible. The dog will attack the mailman if he walks along the path.

How might the mailman manage to deliver the letter safely?

THE GERM JAR

X and Y are two types of germ that exhibit the following predator-and-prey behavior.

X germs: A single X will eat one Y per minute if any are available. Once an X has eaten a Y, the X multiplies to become two Xs.

Y germs: A single Y will double every minute into two Ys, unless it is eaten by an X.

In other words, an X only doubles after devouring a Y, but a Y will double on its own. A scientist has a jar containing 30 Y germs, and places inside it a single X germ.

After how many minutes are there no more Y germs left in the jar?

A famous recreational problem about predation concerns a man thrown into a circular arena with a hungry lion. If the man and the lion have the same maximum running speed, and both have limitless energy, will the lion eventually catch the man or can the man always evade the lion?

The problem is notable not only because it captured the imagination of the mathematical community, but also because for several decades people kept getting it wrong. The German mathematician Richard Rado proposed the problem in 1932. (Like the protagonist of the puzzle, Rado, a Jewish man in Berlin, was also fearing for his life. A year later he fled the Nazi regime to Britain.) Originally, the prevailing view was that the man was doomed, that in a bounded circular arena he would not be able to escape the jaws of the hungry lion. Yet in the 1950s, the Cambridge professor Abram S. Besicovitch discovered a strategy in which the man would, in fact, be able to dodge the lion ad infinitum. The man was saved. While the proof is too involved for this book, the strategy is easy to understand: The man will evade capture if, for each time interval, he runs in a line perpendicular to the line between him and the lion, choosing the direction that keeps him closer to the center point of the circle.

Here's a simpler challenge involving a circular arena and a peckish predator.

$$\overset{\Large(22)}{}$$

THE FOX AND THE DUCK

A duck is floating in the middle of a circular lake. A fox is prowling around the bank. The fox can run four times faster than the duck can paddle, and the fox will always position itself in the best possible spot on the side of the lake to catch the duck.

The duck can only fly from dry land. Is there a way for the duck to reach the side of the lake—and fly to safety—without being caught by the fox?

At first glance, the duck's prospects don't look good.

Let's say that the fox is at the top of the lake, as shown in the image opposite. If the duck decides to paddle in a straight line as fast as it can opposite the direction of the fox, then by the time it gets to the shore the fox will be waiting. We can work this out with some basic geometry. The duck's path is a distance of r, the radius of the lake. The fox's path is a distance of πr. (It travels half the circumference of the lake, and the circumference of a circle is $2\pi r$, where the value of π is approximately 3.14.) The fox thus needs to run a distance that is 3.14 times greater than the distance the duck has to paddle. Since the fox is four times faster than the duck, it will get there first.

The challenge is to plot a path that allows the duck to step onto land at a point the fox can't get to quick enough.

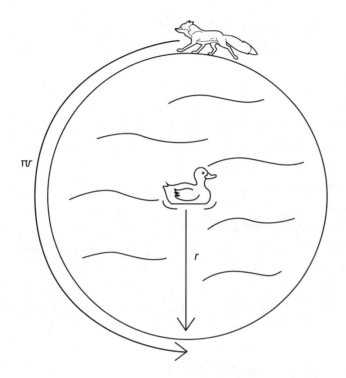

John von Neumann—the speed-solver of the "zig-zagging fly" (problem 10)—is considered, together with the economist Oskar Morgenstern, to be the founding father of game theory, the mathematical analysis of decision-making. The games von Neumann was originally thinking about were parlor games, but as the subject expanded it found wide applications in many fields, such as psychology, philosophy, politics, sociology, and, of course, recreational puzzles. Game theory models the behavior of individuals who interact according to certain rules and aim to "win"—in other words, to get the best possible outcome for themselves. Such as the lions in the following puzzle.

(23)

THE LOGICAL LIONS

Ten lions are in a pen. Their favorite food is sheep. The lions know, however, that any lion that eats a sheep will become drowsy, and is likely to be eaten by another lion if one is nearby. A lion that eats a lion will also become drowsy, and thus also risks being eaten.

A sheep is put in the pen. Each of the lions is desperate to eat it, but will only do so if they are sure they will not be eaten themselves.

What happens to the sheep?

(Extra question: What happens if there are 11 lions in the pen at the start?)

To avoid ambiguity, if the sheep is eaten, it must be eaten by one lion only, and not shared among the pack. We can assume that lions will always act in their best interests and are impeccable logicians.

Pigs are highly intelligent creatures, even the ones that don't have PhDs. In 1979, Basil Baldwin and G. B. Meese, of the Babraham Institute, Cambridge, conducted an experiment involving pigs, as described in the following puzzle. The experiment became famous because it showed that game theory perfectly modeled the animals' behavior.

TWO PIGS IN A BOX

Two pigs live in a box: a bigger, dominant one, and a smaller, subordinate one. The box is arranged such that when a lever is pressed at one end, food is dispensed in a bowl at the other. The distance between the lever and the bowl means that the pig pressing the lever will get to any newly dispensed food second.

Which pig eats better?

After all this talk of food, here's some wine to wash down the final puzzle in this chapter.

TEN RATS AND ONE THOUSAND BOTTLES

You have inherited a collection of one thousand bottles. All the bottles contain wine except one, which contains poison. The only way to discover what's in a bottle is to drink it. If you drink poison, however, you die.

Thankfully, you have 10 rats. If a rat sips poison, or poison mixed with wine, it will die after exactly one hour. If a rat drinks wine, it survives. How do you determine which bottle is poisoned exactly one hour after the first rats are given their first sips?

With unlimited time, 10 rats are plenty to help you discover the poisoned bottle. For example, you could separate the thousand bottles into 10 batches of 100, and allocate each batch to a different rat. Let the rats take a sip from every bottle in their batch. An hour later one rat will be dead, narrowing the options for the poisoned bottle to the 100 bottles in that rat's batch. Separate these 100 into 9 smaller batches, one each for the 9 remaining rats. Again, a dead rat will pinpoint the contaminated batch. Carry on in this way and the rats will soon narrow the options to a single poisoned bottle.

Poisoning rats in series, however, is unnecessarily long-winded. The solution is to poison in parallel, in other words to get the rats to quaff contrasting concoctions concurrently. At least one rat will remain alive during your attempt at mass intoxication.

Dodging death sets us up for the next chapter. It begins with a curious animal, perhaps the best-known puzzler of all.

Tasty teasers

Grueling grids

1) What is the smallest number of straight lines you need to draw across a 3 × 3 grid so that every cell in the grid has at least one of the lines passing through it? The answer is fewer than three. Draw your solution.

2) What is the smallest number of straight lines you need to draw across a 4 × 4 grid so that every cell in the grid has at least one of the lines passing through it? The answer is fewer than four. Draw your solution.

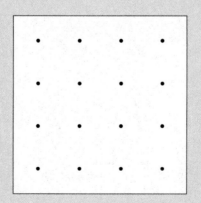

3) Place five stones on the 8 × 8 grid shown below in such a way that every square consisting of nine cells has only one stone on it

The remaining questions all ask you to draw a pattern on a grid of 16 dots, shown below.

4) A polygon is a shape in which each side is a straight line. The H polygon below has 12 sides and the K has 13. Draw a polygon on the grid with 16 sides. (Note: Each side of the polygon must join two dots. Lines cannot overlap. The shape, which need not resemble a letter, must have no gaps in its outline, and each dot can be passed through at most once.)

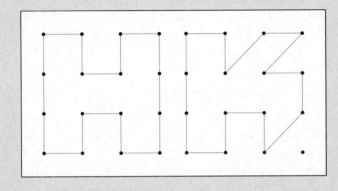

5) Below is a single square made by joining four dots in the grid. Find the other 19 squares that can be made by joining four dots. (Lines connecting the four dots may pass through other dots.)

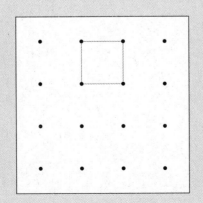

6)   The illustration below shows a way to connect 14 of the dots with lines, such that the angle at every dot is acute (i.e., less than 90 degrees). Find a way to connect all 16 dots so that there is an acute angle at every dot.

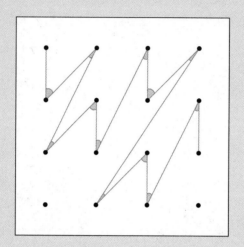

I'm a mathematician, get me out of here

SURVIVAL PROBLEMS

The most famous riddle of the ancient world was also a high-stakes challenge: Get it wrong, and you were eaten alive.

What walks on four legs in the morning, two legs at noon, and three legs in the evening?, asked the Sphinx, the mythical lion-woman, of travelers approaching Thebes. Oedipus replied: "Man, who crawls on all fours as a baby, then walks on two feet as an adult, and then uses a walking stick in old age." Oedipus's reward for solving the puzzle was freedom to enter Thebes, where he would eventually kill his father by accident, marry his mother, and blind himself by sticking pins in his eyes. Maybe he should have given the Sphinx the wrong answer—he would have saved everyone a lot of trouble.

(In some tellings of the Oedipus story, the Sphinx has a second riddle: *There are two sisters: one gives birth to the other and she, in turn, gives birth to the first. Who are the two sisters?* The answer will be revealed in the following pages.)

Jeopardy adds spice to a puzzle. In this chapter you will also use your wits to save your skin. Just as Oedipus faced off against the Sphinx, you will go mano a mano against evil executioners, malevolent monarchs, and perverse prison wardens, each one setting a challenge more fiendish than the last. You—and others—will find yourselves in all manner of apparently impossible pickles: in a dungeon, on a boat, in handcuffs, stranded on distant islands, and incarcerated in some really weird jails. Some of these problems are about escape. Some are about survival. All of them are about having a plan.

FIRE ISLAND

You are marooned on a rectangular-shaped island that stretches 500 m from north to south and 3 km from west to east. The island is surrounded by high cliffs, and is completely covered in dry, combustible forest. The wind is constant and always blows from west to east. On Monday at noon the entire western edge of the island catches fire. The fire, aided by the wind, moves 100 m eastward every hour. If you are trapped by the flames, or if you jump off the island, you will die. In your rucksack you have a compass, a calculator, a penknife, and a copy of the Bible.

How do you survive until Wednesday?

THE BROKEN STEERING WHEEL

You are in a getaway car driving through empty wasteland. Suddenly you realize that the steering wheel is broken: It does not turn right. You can only travel in two directions: straight ahead and left.

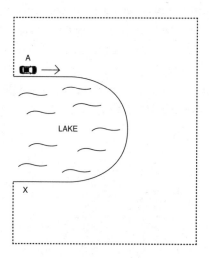

You are at point A, moving in the direction of the arrow. How do you drive to your hideout at X on the other side of the lake without going beyond the demarcated area (shown by the dotted line)? You are not allowed to reverse, get out of the car, or cross the lake.

Puzzles can save your life. That's why they're worth studying, wrote the French mathematician Claude-Gaspard Bachet de Méziriac in 1612, in what is generally regarded as the earliest published book of recreational mathematics, *Problèmes Plaisants et Délectables Qui Se Font Par Les Nombres* (*Amusing and Entertaining Problems That Can Be Had with Numbers*). Dismiss recreational problems at your peril, he says, since they will come in handy in all manner of life-and-death situations.

Consider Titus Flavius Josephus, a Jewish scholar and one of several men hiding in a cave during the Siege of Yodfat in 67 CE. Rather than die at the hands of the enemy, explains Bachet, the men decided to kill themselves. Josephus volunteered a "fair" plan about how they might carry out the collective suicide. The men should stand in a circle, count out a fixed number, kill the person counted out; count out the same number again, kill that person, and so on, until no one was left. Thanks to his mathematical nous, Josephus' magnanimous suggestion was in fact a reverse-engineered guarantee of his own safety. "Josephus chose himself such a good starting position," writes Bachet, "such that if the slaughter continued to the end he would be the last one alive, or maybe he would save a couple of his most trusted friends."

Even though there is no clear evidence that Josephus did indeed use this method to survive the siege, "counting-out" puzzles are called "Josephus problems." They were among the most popular recreational brainteasers of the Renaissance, and the first famous puzzles in which the aim is evading execution. In *Problèmes*, Bachet sets out the most common variation. A boat has 30 passengers. Half of them are to be thrown overboard. If you know which 15 you want to drown, how do you arrange the 30 in a circle such that if every ninth person is thrown overboard, the first 15 counted out are your preferred choices? (The solution is encoded in Bachet's phrase *Mort, tu ne falliras pas, En me livrant le trépas!*: "Death, you will not fail to deliver my demise!" Can you work it out? The answer is in the back.)

Here's a more manageable variation with only 10 on deck.

WALK THE PLANK

Pirates have captured five British and five French sailors. The pirate captain decides that five of the captives must be thrown into the sea. The sailors are positioned in a circle in this order:

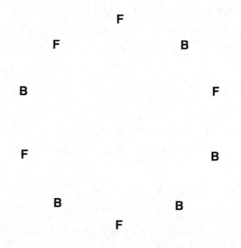

The captain will start with an arbitrary sailor, in position *a*, and count out clockwise every *b*th sailor. He will save the first five counted out, and the others will face a watery death.

In order to save the five British sailors, what are *a* and *b*?

In order to save the five French sailors, what are *a* and *b*?

Let position 1 be the 12 o'clock position occupied by the French sailor, with the other positions counted clockwise from there.

(Note that once a person is counted out they are not included in any subsequent count.)

Death by drowning features in almost all Western Josephus problems. In Japan, however, the problem is told as a fable about hubris. A farmer has a certain number of children with his first wife and a certain number with his second. In order to decide who inherits his wealth, the children are arranged in a circle and every tenth child is counted out until only one—the single heir—is left. One of the mothers starts the counting, aiming for her child to be the winner, but she makes a mistake due to overconfidence, and ends up counting out all her own kids.

Are you still thinking about the Sphinx's second riddle? *There are two sisters: One gives birth to the other and she, in turn, gives birth to the first. Who are the two sisters?* The answer is night and day, which replace each other endlessly as Earth spins around its axis.

THE THREE BOXES

[1] You are locked in a dungeon. By the door are three boxes: one black, one white, and one red. The boxes have the following sentences written on them:

BLACK	WHITE	RED
The key is in this box	The key is not in this box	The key is not in the black box

A sign by the boxes reads: "Of these three statements, at most one is true." If you are only allowed to open one box, which one do you open to find the key?

[2] Let's say you find the key. The door opens, and leads to another dungeon, again containing three boxes:

BLACK	WHITE	RED
The key is not in the white box	The key is not in this box	The key is in this box

A sign by these boxes reads: "Of these three statements, at least one is true and at least one is false." If you are only allowed to open one box, which one do you open to find the key?

From keys in boxes to a box with locks in.

(30)

SAFE PASSAGE

You want to send your beloved a ring using a mail service that is notorious for opening packages and stealing valuables. To ensure the ring arrives safely it must be sent in a lockbox, such as the one shown on the right, which has five holes for padlocks. (A padlock in any of the five holes will lock the box securely.) You and your beloved have five padlocks each. You also have the keys for your own padlocks, but you don't have the keys for each other's padlocks.

If you have unlimited money for postage, how are you able to send your beloved the ring?

Solving a puzzle can be like unlocking a mystery. In these next puzzles, quite literally so.

CRACK THE CODE

A combination lock has a three-digit key. Use the clues below to deduce the code.

6	8	2	One number is correct and correctly placed
6	4	5	One number is correct but wrongly placed
2	0	6	Two numbers are correct but wrongly placed
7	3	8	Nothing is correct
7	8	0	One number is correct but wrongly placed

(32)

GUESS THE PASSWORD

The password to open a door consists of seven digits, no two of which are the same. So 0123456 is a possible password, for example, but 0123455 is not. The door will open if you type in a seven-digit number consisting of seven different digits, and at least one of the digits matches its position in the password. For example, if the password is 0123456, the door will open if you type, say, 0456789, because both your number and the password have a 0 in the first position.

What is the smallest number of attempts you must make to guarantee that the door will open?

(33)

THE SPINNING SWITCHES

You are locked in a cell. On the door is a wheel, illustrated below, which has two identical buttons on it. The buttons are switches that can be pressed on and off, but there is no way to tell whether a switch is on or off.

The door will unlock when both buttons are on. At any stage you can press either one button or both buttons together. Once you have had a try, either the door will open or the wheel will spin, and when it comes to rest you won't know which button is which.

How do you unlock the door in a maximum of three moves?

If that was too easy, try the four-switch version. The wheel now has four buttons, one in each of the four compass directions. The door will open only when all four are on. At any stage you can press one, two, three, or four buttons simultaneously. If the door does not open the wheel will spin, and when it comes to rest you won't know which button is which. What strategy will get you out? (The solution to this one is also in the Answers section.)

You've spent long enough alone. Let's introduce some other people.

PROTECT THE SAFE

The three directors of a bank are suspicious of one another, and agree on a system of locks and keys for the bank's safe, such that:

No single director can open the safe alone.

Any two directors can open the safe by pooling their keys.

What's the smallest number of locks and keys they need to open the safe, and how do they distribute them?

Three mistrustful work colleagues also feature in a lovely puzzle about extracting information without giving anything away. *How do three people find the average of their salaries without any one of them disclosing their actual salaries?* (The solution is in the Answers section.) Here's a similar problem, with extra tattoos.

35

THE SECRET NUMBER

Every gang in your area has a secret number, and the only way gang members can correctly identify fellow members of the same gang is by checking this number.

You're thrown in prison. An inmate comes up to you and claims he's in your gang. You're suspicious. You're not going to reveal your gang's secret number, just in case the inmate is a member of a rival gang, and for the same reason the inmate isn't going to reveal his number to you either.

Another prisoner, Lag, joins the conversation. Lag says that either of you can tell him anything, or ask him anything, and he will answer truthfully, and quietly, so that the other one doesn't hear. (He's also discreet, so he won't listen to any conversations between you and the inmate.)

How do you discover whether you and the inmate are in the same gang, without either of you revealing your number to each other, or even to Lag?

After Claude-Gaspard Bachet's *Problèmes Plaisants et Délectables* in 1612, the next great book of puzzles was *Récréations Mathématiques et Physiques*, a book of math, physics, and magic tricks by the French author Jacques Ozanam. Its second edition, published in 1723, contained the following parlor game.

REMOVING THE HANDCUFFS

Two people are joined together by two pieces of string looped around their wrists, as shown below.

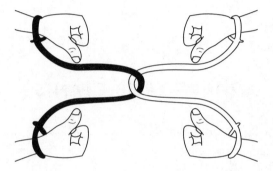

How can the people free themselves without untying or cutting the string?

Try it out yourself with a friend. In fact, why not follow the advice I read in a 1950s magic book: The next time you have a party, divide your guests into couples, join each couple with string handcuffs, and give a prize to the first pair to unlink themselves. It guarantees that couples will "undertake astonishing contortions in fruitless efforts" to set themselves free.

Magic is the art of doing the seemingly impossible, and illusions are often based on mathematical surprises. The trick presented in the handcuffs puzzle is unraveled with *topology*, a type of geometry concerned with the properties of objects that don't change if the objects are stretched or squeezed. To a topologist, all closed loops are the same, whatever the size or material: The metal ring on your finger, for example, is topologically identical to a hula hoop. If two closed loops are interlocked, you cannot free them without breaking one of the loops.

The handcuffs puzzle appears to involve two interlocked closed loops. But it doesn't, and realizing that is the way to the solution.

If you don't have any friends around, here's a topological party trick you can practice on your own.

(37)

THE REVERSIBLE PANTS

Tie a 1 m length of rope between your ankles as shown below. Without cutting or untying the rope, how do you take your pants off and put them back on inside out, with the fly on the front?

A standard challenge in escapology (once your pants are back on correctly) is how to exit a maze. The following puzzle requires you to enter one.

MEGA AREA MAZE

Find the mystery area without using fractions in any of your calculations.

7 cm		

35 cm² 20 cm² 21 cm² 21 cm² 14 cm² 40 cm² 40 cm²

42 cm² 21 cm² 25 cm² 18 cm² 20 cm² 14 cm² 15 cm²

72 cm² 10 cm² 45 cm² 12 cm² 30 cm² 15 cm² 42 cm² 28 cm² 12 cm²

20 cm² 9 cm² 16 cm² 54 cm²

55 cm² 16 cm² ? cm² 40 cm² 10 cm² 32 cm² 36 cm² 15 cm²

21 cm² 18 cm²

20 cm² 27 cm² 21 cm² 16 cm² 16 cm² 9 cm² 12 cm² 21 cm² 24 cm²

15 cm² 36 cm²

25 cm² 10 cm² 45 cm² 56 cm² 20 cm² 20 cm² 15 cm² 15 cm²

40 cm² 12 cm² 16 cm² 21 cm² 27 cm² 45 cm² 42 cm² 30 cm²

36 cm² 36 cm² 10 cm² 35 cm²

All you need to know to solve this puzzle, by the Japanese designer Naoki Inaba, is that the area of a rectangle is equal to the product of its side lengths. I'll start you off. The top left rectangle has an area of 35 cm² and one side of length 7 cm, so we can deduce that its other side is 5 cm long. From this we can deduce that the rectangle of area 20 cm² to the right of the first has sides of length 5 cm and 4 cm, so the rectangle of area 21 cm² below that also has a width of 4 cm.

You will then pass through almost every rectangle in the grid before reaching the mystery square.

Now that you're in a maze, your job is to find a way out.

(39)

ARROW MAZE

The maze shown below is an 8 × 8 grid, and each cell contains an arrow pointing up, down, left, or right. You are in the top left cell, and the only exit from the maze is the right-hand edge of the bottom right cell.

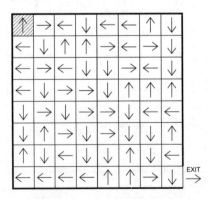

The rules of the maze state that from any cell you must move one cell in the direction of the arrow. In other words, you follow the arrow into an adjacent cell. Once you arrive in a new cell, however, the arrow in the cell you moved from is turned clockwise by 90 degrees. If you arrive in a cell that points out of the grid, you stay in that cell and the arrow is turned 90 degrees clockwise, unless you are in the bottom right cell, with the arrow pointing right, in which case you exit the grid.

Will you ever get out of the grid?

Let's start. The arrow in the top left cell points up. You can't get out, so the arrow in that cell turns 90 degrees clockwise. It's now pointing right. You move to the adjacent cell, where the arrow again points right.

You could carry on like this, following the arrows on the page, to see whether you eventually escape the grid. And if you did, you would quite rightly complain that it was a really frustrating and unsatisfactory puzzle, since you have to remember the new arrow directions of every cell you pass through. My advice is not to follow the arrows or trace any particular path at all.

Instead, consider what would happen if the grid was much, much larger—say a million cells by a million cells, and if you had no idea of the original directions of the arrows. What would happen then?

40

THE TWENTY-FOUR GUARDS

A prison is surrounded by 24 guards arranged in groups of three, as shown in the diagram below, with nine guards on each side.

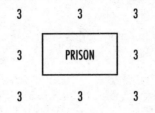

Prison regulations state that *exactly* nine guards must be along each side of the prison at all times. However:

[1] On Monday night, four of the guards sneak off to the bar.

[2] On Tuesday night, the 24 guards are joined by four other guards.

[3] On Wednesday night, the 24 guards are joined by eight other guards.

[4] On Thursday night, the 24 guards are joined by 12 other guards.

[5] On Friday night, six guards go to the cinema.

The guards never break the rules. How do they manage it?

$\left(41\right)$

THE TWO ENVELOPES

You are brought to the king, who is sitting at his desk, by a big log fire. "Today you will learn your fate," he says. "Either you will be executed or you will be freed." He points to two envelopes on his desk. "In one envelope is the word DEATH. In the other is the word PARDON. Choose one of them, open it, and I will act on whichever word is revealed."

As you deliberate which envelope to choose, a guard whispers to you that both envelopes contain the word DEATH.

If he is correct, can you think of a strategy to survive?

$\left(42\right)$

THE MISSING NUMBER

You are locked up in a distant land. The mischievous monarch will only free you if you pass the following numerical test. "I will read aloud every number between 1 and 100 leaving one number out, and your task is to tell me the number I omitted," says the queen. "I will recite my list of 99 numbers in a random order, at a rate of one number every two seconds. I will read each number aloud only once. You have neither pen nor paper, nor any other way of recording what I say."

You don't have a very good memory, so memorizing the list is beyond your capabilities. But you are good at arithmetic. What's a simple strategy for finding the missing number?

You find the missing number. Now the queen—who has a penchant for puzzles involving the numbers from 1 to 100—pits you against a fellow prisoner.

(43)

THE ONE HUNDRED CHALLENGE

You are to play a game against your cellmate, in which you both take turns to say a number between one and 10. The loser is whoever brings the overall total to 100 (or more). Your cellmate starts.

"Eight," he says.

"Three," you reply.

"Four," he says.

What strategy guarantees that you will win?

Islands that appear in puzzles often have inhabitants with peculiar behavioral traits. Here we visit one that has been on the tourist trail since at least the 1950s.

(44)

THE FORK IN THE ROAD

You are on an island inhabited by two tribes. The members of one tribe always tell the truth. The members of the other tribe always lie. You come to a fork in the road. One branch leads to the airport, the other to the beach. You spot a local standing nearby, but you cannot tell which tribe she belongs to.

You must ask the local a *single* question to discover which branch leads to the airport.

What do you ask her?

The question seems to ask the impossible because, intuitively, one assumes that people who tell the truth and people who lie always give contradictory answers to the same questions. For example, if you pointed at a road and asked the local whether it was the correct route to the airport, her answer would depend on which tribe she belonged to. A truth-teller would say the opposite to a liar. If the former said yes, the latter would say no, and vice versa.

Yet there are questions that force liars to tell the truth. The ruse is to get the liar to lie about their own lie. In other words, ask the local a question about how she would answer a question.

$$\left(45\right)$$

BISH AND BOSH

You are on an island inhabited by two tribes. The members of one tribe always tell the truth. The members of the other tribe always lie. You come to a fork in the road. One branch leads to the airport, the other to the beach. You spot a local standing nearby, but you cannot tell which tribe he belongs to.

You want to ask him which branch leads to the airport, but your plight is complicated by the fact that even though the inhabitants of the island understand English, they cannot speak it. Instead, they reply to all English questions with the words "bish" or "bosh." You know that these are native words for "yes" and "no," but you don't know which is "yes" and which is "no."

What question can you ask the local that will reveal which route goes to the airport?

Just as in the previous question, you are not trying to work out whether the local is a liar or not, nor are you trying to work out what "bish" or "bosh" mean. All you want to do is leave the goddamn island!

Raymond Smullyan (1919–2017) was the twentieth century's most brilliant and prolific inventor of logic puzzles, many of which were set on islands containing truth-tellers and liars. The next puzzle is one of his. It's an example of what he calls "coercive logic," since the idea is to force a person, using logic, to do the opposite of what they want to do.

THE LAST REQUEST

You are due to be executed tomorrow. The executioner asks if you have a last request.

"I would like to ask you a question," you say. "All I ask is that you answer it truthfully by saying either 'yes' or 'no.'"

The executioner is relieved, and rather surprised, to discover that the request is so straightforward. He promises to answer the question truthfully.

[1] What question can you ask the executioner that forces him to answer "yes" and spare your life?

[2] What question can you ask that forces the executioner to answer "no" and spare your life?

The solutions to the fork-in-the-road problems relied on the fact that a liar will inadvertently tell the truth when lying about his own lie. The executioner problem, on the other hand, is about the inadvertent consequences of always telling the truth.

I'm going to show you an answer to [1] in the next few paragraphs, so look away now if you want to attack this problem with no help.

You probably didn't look away. This type of logic problem is so brain-twisting it is hard to know where to start. I've included this one because the solution is so clever and elegant that I hope you will derive pleasure from it whether or not you figure it out.

A valid solution to [1] is: *"Will you either answer 'no' to this question or will you spare my life?"*

If you were to ask this question, the executioner would have no alternative but to answer "yes" and spare your life. Let's break it down to understand why. The question is asking whether one of these two statements holds:

[1] The executioner will answer "no" to the question.

[2] The executioner will spare your life.

If the executioner answers "no"—that is, if none of these alternatives hold—then the executioner is not being truthful, since one of these alternatives does hold, namely the first one. The executioner has contradicted himself. The response cannot truthfully be "no," so the executioner is forced to answer "yes."

But for the executioner to answer "yes," one of the alternatives must be true. It cannot be the first one, since the executioner has not answered "no." It must then be the second one, so the executioner has agreed to spare your life. If your brain is still in working order, try part [2].

Puzzles involving the playful interplay of true and false statements date from the nineteenth century. They have become particularly relevant in the past few decades with the growth of computer science. All programming languages rely, at a basic level, on the logic of truth values.

At the turn of the twenty-first century a puzzle in a computer science PhD thesis caused huge excitement. Not only did it reveal an astonishing result, it also had important practical applications. The puzzle is a wonderful example of a simple problem at the cutting edge of research that captured the imagination of the entire mathematical community. Before we get there, though, a warm-up puzzle.

THE RED AND BLUE HATS

Two prisoners are in a cell. The warden announces that they will play a game. A red or blue hat will be placed on each of their heads, with the color of each hat determined by the flip of a coin. Each prisoner will be able to see the other one's hat but not their own.

Once the prisoners see each other's hats, they must guess the color of their own hat. They will both be freed if at least one of them guesses correctly.

The prisoners are allowed to discuss a strategy before the game begins, but once the hats are on, no communication of any sort is allowed.

What strategy guarantees their freedom?

If the game involved only one prisoner guessing the color of his hat, it would, of course, be an impossible challenge. The chance of that prisoner guessing correctly is 50 percent, since there is an equal chance of it being either red or blue. Yet once you add a second player, the chance of at least one correct guess rises to 100 percent, with the right strategy.

In 1998 Todd Ebert, a computer science graduate student at the University of California, included a version of the previous puzzle in his PhD thesis. He upped the number of prisoners to three. Within a few years his problem had been discussed around the world, and the buzz surrounding it was even reported in the pages of *The New York Times*. I'll include it here as an extra puzzle.

The setup is the same: *Three prisoners are in a cell. The warden announces that they will play a game. A red or blue hat will be placed on each of their heads, with the color of each hat determined by the flip of a coin. The prisoners will be able to see one another's hats but not their own.*

Here's the twist: *Once they see each other's hats, at least one prisoner must guess the color of his own hat. If any guesses are wrong, all three prisoners will be killed.*

In other words, up to two prisoners can avoid guessing their hat color by staying silent. Yet at least one of them needs to guess, and needs to guess correctly. Only if there are no incorrect guesses will they all survive.

As before, the prisoners are allowed to discuss what to do beforehand, but once the hats are on no communication is allowed. One obvious strategy would be for the prisoners to designate one person to always guess, and for the other two to always stay silent. This strategy gets you a 50 percent chance of survival, since the designated player will guess correctly half the time. Indeed, on first hearing the problem most mathematicians assumed that it was impossible to improve on 50 percent.

Amazingly, however, there is a strategy that guarantees a survival rate of 75 percent. (The solution is in the back.) If you play the game with 16 prisoners the result is even more stunning: The odds of survival are more than 90 percent. This puzzle was posted on chat rooms and newsgroups and became one of the first viral puzzles of the internet age.

Fascinating and powerful ideas from computer science can be brilliantly illuminated with a good puzzle. The generalized solution to the hats problem, for example, involves the mathematics of Hamming codes, a type of error-correcting code developed for telecommunications in the 1950s. The Boyer-Moore majority vote algorithm, meanwhile, is an ingenious way to remember information with barely any computer memory, and is the solution to the following teaser.

(48)

THE MAJORITY REPORT

A prison warden reads out a long list of names. Some names are read out more than once. In fact, one of the names appears in the list more often than all the other names combined. (In other words, more than 50 percent of the time a name is called out, it's this name.) If you can identify this "majority name" you will be freed.

You are not allowed a pencil and paper, so you have no way of tallying all the names. You have also suffered a knock on the head, which has resulted in an inability to keep more than one name in your working memory at any one time. That is to say, on hearing a name you can choose to remember it, but if you do so then you instantly forget the name that was in your memory before. Or you can also choose *not* to remember it, thus retaining the name that was in your memory before.

The prison warden allows you to use a single clicker-counter. It is set to 0. You can click up the numbers, and down the numbers, at will.

What strategy guarantees that you identify the majority name? You can assume that once you have decided on a strategy, you will not forget it as the warden begins her recital.

If there were only two different names on the list, say Smith and Jones, one strategy that would work would be to click the counter upward on hearing Smith, and downward on hearing Jones. Once all the names had been read out, the clicker would display a positive number if Smith was said more times, and a minus count if it was Jones.

With three or more names, the strategy is a little more complicated. You can't just count up for one name and down for another name, since that leaves other names unaccounted for. At each step in the recital you have three pieces of current information: the name you hear, the name in your memory (which may or may not be different from the one you just heard) and the reading on the clicker. How do you combine these three pieces of information to find the name that is read out more than half the time? It is incredible to discover what can be "remembered" with almost no memory at all.

Puzzles that originated in computer science are often about information: storing it, as above, and communicating it. Prisons are convenient venues for this type of puzzle, because incarceration means restricted access to information. In the two final problems in this chapter, a group of prisoners must think up a strategy for a challenge in which the restrictions on communication are so comically severe the thought of even meeting the challenge is mind-boggling.

(49)

THE ROOM WITH THE LAMP

A room in a prison has a lamp. The lamp is either on or off, and can only be switched on or off by someone in the room.

The prison warden decides that every day he will select one of 23 prisoners to go into the lamp room, after which they will return to their individual cells. The warden's selection is random, meaning that he might choose the same prisoner to go into the room several days in a row, or months might pass before a given prisoner is chosen. In the long run, though, every prisoner will go into the lamp room the same number of times as every other prisoner.

The warden tells the prisoners that he will free them all as soon as one of them says "we have all visited the lamp room"—provided this statement is correct. If a prisoner makes this statement, but at least one prisoner hasn't visited the room, the warden will have all of them killed.

Before the warden chooses his first prisoner, all of the prisoners are brought together to discuss a strategy. Once they decide on their plan, they will return to their cells and will no longer be able to communicate with each other from there. The only way they can send each other messages is via the lamp: They can either leave it on or off. We can assume that no one else goes in the room apart from the prisoners, and that the lamp is always in working order.

Can you think of a strategy that will allow one of the prisoners to say, with 100 percent certainty, "we have all visited the lamp room"?

Note that we do not know whether the lamp is on or off to begin with. To get you on your way, let's simplify the problem. Assume, for a moment, that the lamp is off at the start and that there are just two prisoners, A and B.

The only way that the first prisoner to enter the room can let the other one know he has been there is by turning the switch to "on." Let's say A is the first prisoner in the room. He sees the lamp is off, and he turns it on. If he's chosen again he leaves the lamp on, and continues to do so if he makes further visits. When B eventually enters the room, he will know that A has been there, since the lamp will be lit and only A could have turned it on. B can say with 100 percent certainty that "we have all visited the room." With two prisoners, therefore, the rule for what you do to the lamp can be summarized as:

If it is off, turn it on.

If it is on, leave it on.

Now try to solve the problem with a third prisoner, still with lamp being off from the start. This strategy can be extended to any number of prisoners. The step after that is to solve the puzzle when the prisoners *don't* know the starting position.

The final puzzle in this chapter also involves a prison and a room containing an object that prisoners must interact with in a clever way. Yet rather than leaving messages for one another, they are looking for hidden information.

(50)

THE ONE HUNDRED DRAWERS

A cabinet with a hundred drawers, each numbered from 1 to 100, stands in an otherwise empty room in a prison. The warden enters the room, writes down the names of 100 prisoners on separate pieces of paper, and distributes the pieces of paper randomly in the cabinet so that each drawer contains the name of one prisoner.

The warden exits the room and assembles the 100 prisoners whose names she wrote on the slips of paper. She tells them that they will each be let into the cabinet room, one after the other: Once there, they are permitted to open 50 drawers and look at the names on the pieces of paper inside. She adds that if every single prisoner opens the drawer containing his name, everyone will be freed. But if at least one prisoner doesn't open the drawer with his name in it, everyone will be killed.

The prisoners are allowed to think up a strategy before they go into the cabinet room, but once the first prisoner goes in they are not allowed to communicate in any way. They cannot leave messages in the room, nor tell other prisoners what they saw once they leave the room.

Can you find the strategy that gives a chance of survival of more than 30 percent?

Of all the mathematical surprises in this book, this result is perhaps the most staggering. If a prisoner chooses 50 out of 100 boxes at random, the chance that one of these boxes contains his name is 50 out of 100, or 50 percent. If all 100 prisoners were to choose 50 boxes at random, the probability of everyone choosing their name would be:

50 percent of 50 percent of 50 percent ... (repeated 100 times) ... which is: 0.0000000000000000000000000008 percent

Yet a strategy exists that improves the chances of *every* prisoner finding his name by a factor of more than 100 octillion.

The strategy is very simple to describe, and relies on a very interesting piece of mathematics that is fully explained in the back of the book. Even if you can't prove that it gives the prisoners more than a 30 percent chance of survival, have a guess as to what it might be. You will be genuinely astounded at how a collective decision can beat the odds.

Tasty teasers

Riotous riddles

Each of these puzzles relies on the solver's making a mistaken assumption at first.

1) Two identical children born on the same day to the same mother and father are not twins.

   How is this possible?

2) A certain man had great grandchildren, yet none of his grandchildren had any children.

   How is this possible?

3) You are on a plane, a mile above sea level. Huge mountains lie directly ahead. The pilot does not change his course, speed, or elevation. Yet you survive.

   How is this possible?

4) A woman has a bucket of water in her hands. She turns it upside down, but the bucket stays full. There is no bucket lid, the water is in liquid form, and she is not relying on centrifugal force.

   How is this possible?

5) A clean-shaven teenager told his parents he was going to a party, and would be back before sunrise. He got back before sunrise, and had a fully-grown beard!

   How is this possible?

6) A boxer left a contest victorious, winning the national championship. Even though his trainer took all the money, the boxer was quite happy.

How is this possible?

7) Identical twins Milly and Molly always wear the same clothes. One day I saw one of them and I shouted hello. As soon as she turned to me I knew it was Milly.

How is this possible?

8) A woman working in an office is fired from her job. The following day she shows up at the same office, where she is welcomed.

How is this possible?

9) My neighbor is 92 years old and frail. One day I invited her over to do something I was not able to do. She was able to carry out the task, even though she has no skills that I do not have.

How is this possible?

10) One day a man saw the sun rise in the west.

How is this possible?

Cakes, cubes, and a cobbler's knife

GEOMETRY PROBLEMS

As professor of minerology at Geneva University in the early eighteenth century, Louis Albert Necker spent many hours staring at diagrams of crystalline forms. He noticed when examining these figures that they would often suffer a "sudden and involuntary change in [their] apparent position." This phenomenon is clearly seen in the image below, now known as the "Necker cube." A flat figure comprising 12 lines, it could equally be a cube seen from above (in which the bottom left square is the front), or a cube seen from below (in which the top right square is the front). Since there are no clues as to the depth of the figure, its orientation when visualized in three dimensions is ambiguous. What's fascinating about the Necker cube is that when seeing it one way, it is impossible to see it in the other way, and the perceived orientation can flip suddenly between views.

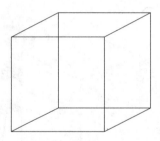

The Necker cube is an optical illusion. For me, it's also a metaphor for the joys of geometric puzzles. The "sudden and involuntary change" in its perceived orientation is analogous to the flash of insight that this type of puzzle requires. We will see many geometric puzzles in this chapter. You will stare at them, searching for clues. You will feel you are getting nowhere, when, all of a sudden, a magical sensation occurs: the thrill of realizing you've been staring at the answer all along. Your eye just flipped between orientations. It's a transcendent feeling. The answer was always there, hiding in plain sight. Once you look at a geometric puzzle the right way, the answer often jumps off the page.

(51)

THE BOX OF CALISSONS

A *calisson* is a French candy made from a candied fruit marzipan topped with icing. Let's assume its diamond shape is made from two equilateral triangles meeting along an edge. When you pack calissons in a hexagonal box, they fit in one of three ways: horizontally ⬦, sloping left ▱ or sloping right ▱. The image below shows the aerial view of one possible arrangement.

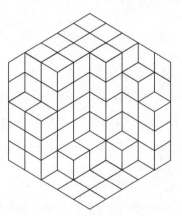

Without counting them, show that the number of calissons in each orientation is the same.

Problems are all the more appetizing when set in the context of pâtisserie. The cake counter, in fact, offers up some delicious brain food. And to think you doubted that the puzzles in this book had any practical applications.

THE NIBBLED CAKE

Someone ate a thin rectangular slice of your cake. Here's an aerial view.

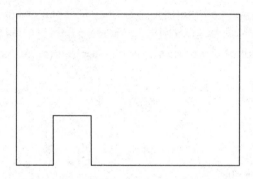

How would you slice the cake with a single straight cut that divides what's left into two equally sized portions?

In other words, draw a single line that divides the shape above into two parts of equal area. Standing the cake on its side and cutting it into a top half and a bottom half doesn't count. That's cheating.

Mmmm, these cake puzzles are so tasty!

CAKE FOR FIVE

How do you cut a square cake into five portions of equal size? Each slice must be "slice-like," meaning that the knife cuts vertically through the cake and the tip of each slice is at the cake's center. You have no ruler or tape measure, but you can use the horizontal grid shown here.

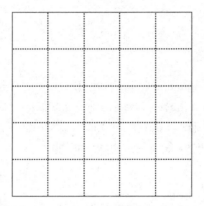

The previous two questions concerned equitable cutting—that is, making sure that every slice contains an equal amount of cake. In real life, we take it for granted that when we share a cake every slice must be the same. In fact, the implicit assumption of fairness has made cake-cutting the favored example in the field of "fair division," an area of math, economics, and game theory that analyzes strategies for dividing things up. Geometers, on the other hand, are not always interested in slicing pastries into equal portions. Sometimes they want to know how to cut an object into the most pieces with the fewest number of cuts.

SHARE THE DOUGHNUT

If you slice a doughnut, like the one above, with one straight-line cut, you'll get two pieces wherever you make the cut.

How do you slice a doughnut with two straight-line cuts to make five separate pieces, and with three straight-line cuts to make nine separate pieces?

I mentioned the joy of geometric puzzles. Let's not forget the pain. Puzzles that require you to cut an object into as many pieces as possible are surprisingly hard. You can find yourself staring at the page for what seems like an eternity and making no progress. These puzzles are almost twice as painful because they look like they should be easy. But there's no trickery involved.

A STAR IS BORN

The five-pointed star shown below contains five "disjoint" triangles—they don't overlap, nor is any one triangle contained inside another. Draw two straight lines across the five-pointed star in such a way that the resulting drawing contains 10 disjoint triangles. I've included an extra star for practice, since it's unlikely you'll get it the first time.

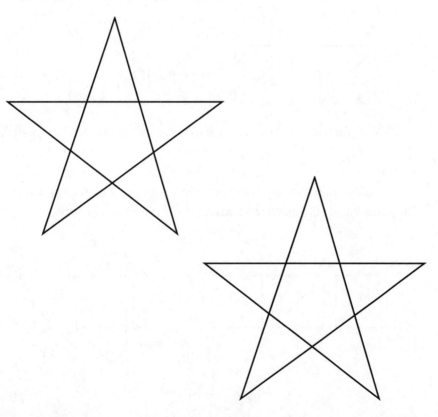

The Greek historian Herodotus wrote that the study of geometry originated when ancient Egyptian tax inspectors were dispatched with rope to measure fields flooded by the Nile. Members of other traditional professions, however, were also practicing geometry at that time, at least informally. Carpenters and tailors, for example, are always faced with the problem of how to cut a piece of wood, or cloth, in the most efficient way possible. How, for example, do you cut up two identically sized squares of material to make the biggest possible square?

Here's the most efficient way:

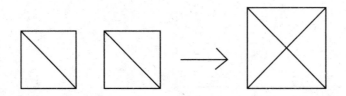

What if you had *three* identically sized squares of material and wanted to make the biggest possible square? Here's a nifty solution from the tenth century that divides the squares into nine pieces. (First cut two of the squares along the diagonals. Then position the triangles as below right, cut the parts outside the dotted square and insert them into the gaps.)

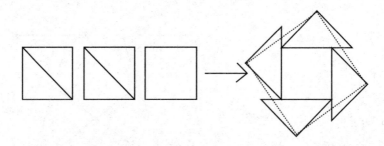

The Persian astronomer Abul Wefa described this method in *On Those Parts of Geometry Needed by Craftsmen*, a treatise he wrote out of exasperation at the lack of communication between artisans and geometers. When it came to cutting up shapes and fitting them back together, the artisans made erroneous deductions, while the geometers had no experience in actually cutting real things. Perhaps the theoreticians and the practitioners should talk!

"Dissection puzzle" is the term for a problem in which a shape is cut up into pieces and reassembled into another shape. The following dissection puzzle dates from the sixteenth century.

(56)

SQUARING THE RECTANGLE

A tailor has a rectangular piece of fabric that measures 16 cm by 25 cm. How does he cut the fabric into two pieces that can fit together to make a square?

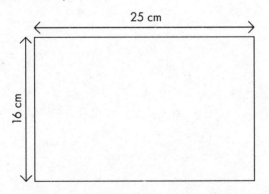

25 cm

16 cm

Readers who bought a 701 series IBM ThinkPad in 1995 will have no problem solving this puzzle.

In the nineteenth century, both academic and recreational mathematicians had fun with dissections. The German mathematician David Hilbert proved that any polygon (a shape with straight edges) can be transformed into any other polygon of equal area by cutting it into a finite number of pieces and reassembling them. In 1900 he published a list of 23 important open problems, which was hugely influential in setting a direction for twentieth-century mathematics. The third question on the list concerned the dissection of polyhedra (three-dimensional solids with flat sides). Given two polyhedra of equal volume, is it possible to cut the first into a finitely large number of polyhedral pieces that can be reassembled to yield the second? That same year, someone proved that it's impossible, making it the first of Hilbert's problems to be resolved.

Dissection puzzles flourished in the late nineteenth and early twentieth centuries. For Sam Loyd and Henry Dudeney, the most prolific puzzle setters of the age, they were the tools of the trade. Loyd, an American, liked to bring them to life with quirky setups. "Speaking about modes of conveyance in China . . ." he wrote as an introduction to the following problem.

THE SEDAN CHAIR

Cut the shape shown below—which looks like a cross-section of a sedan chair—into two pieces that will fit together to form a square.

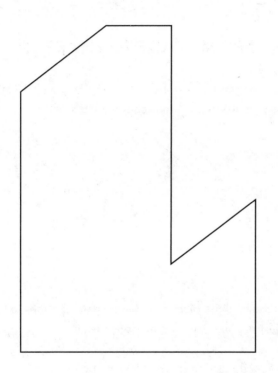

The setup for another of Loyd's puzzles began: "During a recent visit to the Crescent City Whist and Chess Club, my attention was called to the curious feature of a red spade which appears in one of the windows of the main reception room." Curious indeed.

(58)

FROM SPADE TO HEART

Imagine that the spade shown below is red. Can you cut it into three pieces to make a heart, thus changing its suit?

You must use all three pieces, and none must overlap—otherwise you could just snip the stem and be done with it.

I like the following dissection puzzle, devised by the French writer Pierre Berloquin, because we must transform a shape with curved edges into a shape with only straight ones.

THE BROKEN VASE

With two straight-line cuts, divide the vase into three pieces that can be reassembled to form a square.

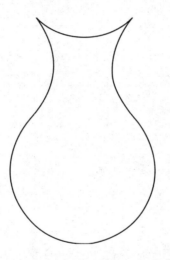

Henry Dudeney, a British puzzler, was particularly innovative and ingenious in his dissection puzzles. Hilbert had showed that any polygon could be transformed into another, but his proof relied on dividing the original polygon into a potentially huge number of smaller triangles. Dudeney, on the other hand, valued elegance, and developed an extraordinary capacity to achieve beautiful transformations with as few cuts as possible.

We saw earlier that Abul Wefa needed nine pieces to transform three small squares into a big square. Dudeney did it with six.

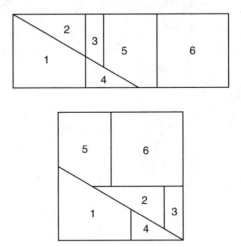

Dudeney's most famous geometric discovery was the four-piece dissection that transforms an equilateral triangle into a square, shown below. If you link the pieces by their corners, as if they're connected by hinges, folding them one way produces the triangle and folding them the other way produces the square. Dudeney was so proud of his discovery that he made a physical model of it with mahogany pieces and brass hinges, which he presented at a meeting of the Royal Society in 1905.

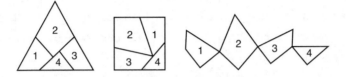

Dudeney's clever dissection puzzles inspired huge interest in this kind of problem throughout the twentieth century. Finding a "minimal dissection"— that is, a dissection from one shape to another that uses the fewest number of pieces—was a perfect recreational challenge, since it required no deep knowledge of geometry. There is no general procedure to find a minimal dissection. What you need is creativity, intuition, and patience. Before the computer age, an enthusiastic amateur might be able to outperform a professional, and many did.

The next few puzzles are also about cutting up shapes in interesting ways.

SQUARING THE SQUARE

The image below shows how to divide a square into four smaller squares.

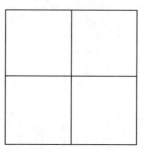

Show how to divide a square into:
   [1] Six smaller squares.
   [2] Seven smaller squares.
   [3] Eight smaller squares.
   In each case the squares can be of different sizes.

That was the warm-up. Now get out the scissors. The next puzzle is from Henry Dudeney's *Amusements in Mathematics* (1917).

MRS. PERKINS'S QUILT

The patchwork quilt shown below is made up of 169 square patches. Divide the quilt into the smallest possible number of squares by cutting only along the sides of patches.

To put it another way, you're trying to find the smallest number of square pieces that can be sewn together to make the quilt.

Mrs. Perkins's Quilt was the first appearance in mathematical literature of the concept of a "squared square"—that is, a big square cut up into smaller ones. Like many of Dudeney's puzzles, it captured the interest of academics.

In the solution to Mrs. Perkins's Quilt, some of the squares are the same size. In the 1930s, mathematicians at Trinity College, Cambridge, set out to find a "perfect" squared square: a square divided into other squares, each of which is a different size. (A group of Polish mathematicians also looked into the problem around the same time.) In 1939, the German mathematician Roland Sprague was the first to publish a solution: a 4205 × 4205 square divided into 55 smaller squares with side lengths of different whole numbers.

For the Dutch computer scientist A. J. W. Duijvestijn, finding squared squares was a lifelong obsession, the subject of his 1962 thesis and of much work over subsequent decades. He wanted to find the smallest possible "simple" perfect squared square, meaning one in which no subset of the squares fits together in a rectangle or square. In 1978 his computer discovered the 112 × 112 squared square shown below, which is made up of only 21 squares. It is the smallest possible simple perfect squared square, and one of the most famous images in mathematics. (The number in each square describes its side length.)

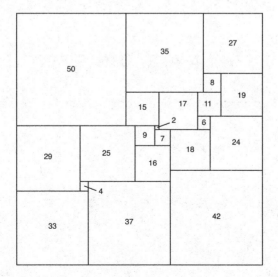

An uncontroversial observation from the image in problem 60 is that four identical squares placed together make a (bigger) square. Likewise, four identical Ls can be placed together to make a (bigger) L, as illustrated below. Shapes that can be placed with identical copies of themselves to make a larger version of the same shape are called "reptiles," because they are tiles that replicate. Equivalently, reptiles are also shapes that can be subdivided into smaller, identical versions of themselves.

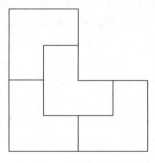

THE SPHINX AND OTHER REPTILES

Divide each of the following shapes into four smaller copies of itself.

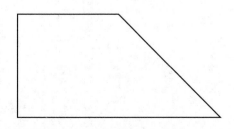

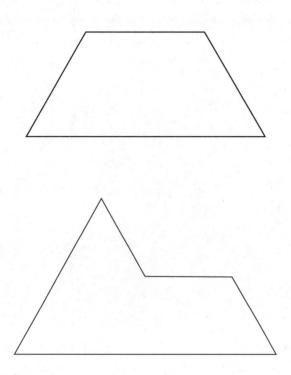

If it helps, the second and third shapes are composed on an underlying grid of equilateral triangles. The third shape is known as a "sphinx," a term coined by T. H. O'Beirne, the *New Scientist*'s puzzle columnist in the early 1960s.

A reptile will tile a flat surface without gaps or overlaps, since you can arrange the tiles to make larger and larger versions of the basic shape. Here's another type of reptile that will also tile a flat surface with no gaps or overlaps (although it will not replicate the original shape in a larger form).

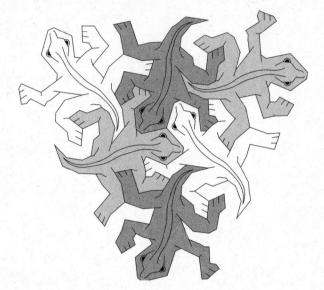

The lizard is based on a tessellation in *Reptiles*, a 1943 lithograph by the Dutch artist M. C. Escher. (A tessellation occurs when a single tile fits perfectly with identical tiles.) Escher created many tessellating tiles in the shapes of living creatures. His interlocking reptiles, fish, and birds make up some of the most recognizable mathematical art of the twentieth century: striking, playful, and utterly ingenious. To create a single tile that looks like a convincing representation of an animal, and which also fits together with identical tiles to leave no gaps or overlaps, is a hard challenge. Escher inspired many others to try the same. The number-one tessellation artist working today is the Frenchman Alain Nicolas, who created the tiles in the following puzzle.

ALAIN'S AMAZING ANIMALS

Divide the following shapes into the number of pieces indicated.
The pieces are in the shapes of creatures.

[1] Two identical pieces.

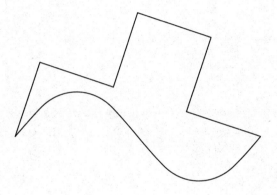

[2] Two identical pieces, one of which is flipped over.

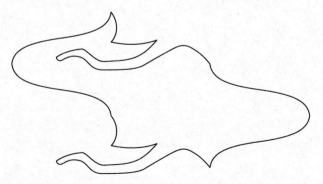

[3] Three identical pieces, one of which is flipped over.

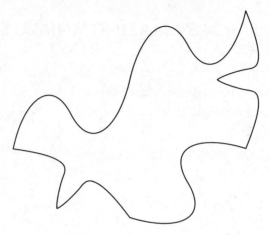

One particularly addictive type of geometric puzzle shows a shape, or shapes, and asks, "What's the area?" Here are two such problems, one with squares and one with a triangle.

THE OVERLAPPING SQUARES

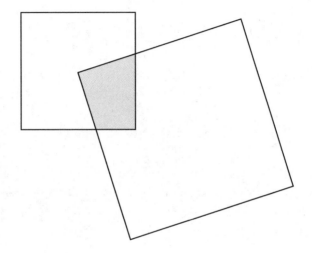

The small square has a side of length 2, and the large square has a side of length 3. The leftmost vertex of the large square is at the center of the small square. The side of the large square cuts the side of the small one two-thirds of the way along.

What's the shaded area?

(65)

THE CUT-UP TRIANGLE

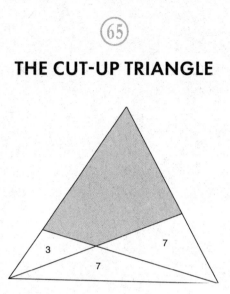

This triangle is divided into four parts, such that the three labeled parts have areas of 3, 7, and 7.

What's the shaded area?

For those of you who have forgotten, the area of a triangle is equal to half the base times the height, where the height is measured perpendicular to the base.

"What's the area" puzzles are both ancient and hypermodern. Ancient because all you need are the rules of geometry set down by the Greeks more than two thousand years ago. Modern because they are eye-catching and shareable, perfect puzzles for the internet age.

Indeed, Catriona Shearer, a math teacher at a school in Essex, has gained thousands of Twitter followers by posting beautiful problems of this type, colored in with felt-tip pens. "I really like puzzles where a bit of clever thinking can sidestep a whole page of algebra," she says. "Plus I enjoy coloring them in." Here are two. The first one concerns circles, so you need to know that the area of a circle is πr^2, where r is the circle's radius. An *arbelos*, from the Greek word for "cobbler's knife," is an area bounded by three semicircles, as shown below.

66

CATRIONA'S ARBELOS

The vertical line, of length 2, is perpendicular to the bases of the three semicircles.

What's the total shaded area?

(67)

CATRIONA'S CROSS

All four shaded triangles are equilateral. What fraction of the rectangle do they cover?

The next question asks you to find an angle, rather than an area. For the remaining problems in this chapter we're in three dimensions. Hold on tight!

CUBE ANGLE

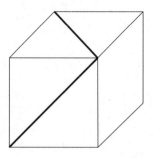

What angle is made by the two bold lines on the sides of the cube?

Prince Rupert of the Rhine is remembered by historians for his dog, and by mathematicians for his cube. A Royalist commander during the English Civil War, he was accompanied in battle by a large white poodle that his enemies, the Parliamentarians, believed had supernatural powers.

When the war was over he became a noted artist and gentleman scientist, helping found the Royal Society. He was also the first person to notice a fascinating property of the cube that defies our geometric intuition. Take two identically sized cubes made of wood. It's possible to cut a hole in one of the cubes and slide the other cube through it.

The anthropophagic ability of a cube to eat itself arises because a cube's width varies depending on where you measure it. For example, place a cube flat on a table. If you make a horizontal slice, the cross section is a square. However, if you balance the cube on one of its vertices, so that the opposing vertex is vertically above it, a horizontal slice through the middle of the cube produces a hexagon. (It's hard to visualize but consider it this way: When the cube is balancing on a point, a horizontal slice through the middle must cut *every* face. There are six faces, so the cross section must have six sides, and because of the symmetry of a cube each side must have the same length.)

The hexagonal cut can also be achieved by a slice that goes through the midpoints of six edges, as shown below.

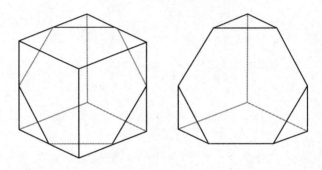

The square cross section of a cube has a smaller area than this hexagonal cross section does. In fact, the square cross section can be made to fit entirely inside the hexagonal cross section. In other words, if you make a hexagonal slice through a wooden 1 m cube and then drill a 1 m square-shaped hole into the slice, you will be left with a thin wooden ring that another 1 m cube can slide through.

Prince Rupert's observation led to a further question: *What is the largest possible cube that can fit through a hole in another cube with a side length of 1 unit?* The problem was not solved for another hundred years, when the Dutch mathematician Pieter Nieuwland showed that a cube with a side length of 1.06 (to two decimal places) can fit through a cube with side length 1. This cube of side length 1.06 is known as "Prince Rupert's cube." (In this optimal case, Prince Rupert's cube does not go through the holed-out cube at an angle perpendicular to the hexagonal slice.)

The next question is about hexagonally slicing a particularly interesting cube called the "Menger sponge," a fractal object first described by the Austrian-American mathematician Karl Menger in 1926. The Menger sponge is a cube with smaller cubes extracted from it, and it is constructed as follows: Step A: Take a cube. Step B: Divide it into 27 smaller "sub-cubes," so it looks just like a Rubik's Cube. Step C: Remove the middle sub-cube in each side, as well as the sub-cube at the center of the original cube, so that if you looked through any hole you would see right through it. Step D: Repeat steps A to C for each of the remaining sub-cubes—that is, imagine that each sub-cube is made from 27 even smaller cubes, and remove the middle one from each side as well as the central one.

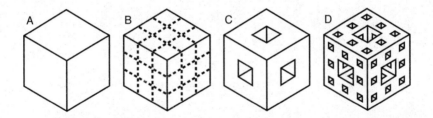

If you repeat steps for A to C for each of these 27 sub-sub-cubes, you get the cube illustrated opposite. (It's called a "level three" Menger Sponge since the iterative process has been carried out three times. You could, if you wanted, carry on the process ad infinitum, on smaller and smaller sub-cubes.)

THE MENGER SLICE

Here is a Menger sponge. If you cut the object in two with a diagonal slice, what does the hexagonal cross-section look like?

Before you look in the back, have a go at sketching what you think you might see, but don't feel bad if you're finding it hard. To deduce the pattern requires phenomenal levels of spatial intuition. I've included the problem not to show you up but to give you a thrill. The geometer who showed me this problem said that seeing the solution was the biggest "wow" he has ever experienced in mathematics.

Prince Rupert was concerned with pushing a cube through a square hole. Now we're concerned not just with square holes but with circular and triangular ones too.

THE PECULIAR PEG

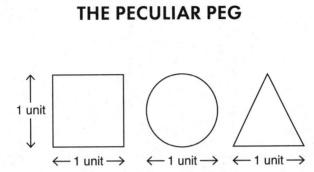

Draw a three-dimensional picture of a solid object that will pass through square, circular, and triangular holes, as illustrated above. The object must touch every point on the inside of each hole as it passes through. In other words, the object has three cross sections that are the same shape and size as these three holes.

Those readers who already know of one solution to this problem, should draw a different object. Or does only one object fit the bill?

My final problem about visualization in three dimensions flummoxed a panel of more than a dozen American college professors. They approved it for use in an aptitude test given in 1980 to 1.3 million high school pupils.

THE TWO PYRAMIDS

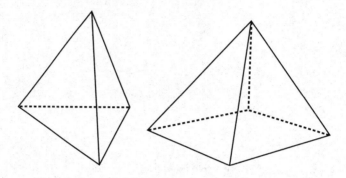

The faces of the two pyramids shown above are equilateral triangles of the same size. One of the pyramids has a triangular base and the other a square base. If the pyramids are glued together so that two triangular faces perfectly coincide, how many exposed faces would the resulting solid have?

(a) 5   (b) 6   (c) 7   (d) 8   (e) 9

The answer given by the examination board was (c), the wrong answer. The board assumed that if one pyramid has 4 sides and the other 5 sides, when you stick them together they each lose a side. Actually, that's not the case. The error was only picked up when one of the students who took the test, Daniel Lowen, aged 17, got his results. After the original exam, Lowen had made a physical model of the two pyramids, which confirmed the answer he had written down. However, when the results were released he discovered he'd got that question wrong. His father—an engineer working on the space shuttle; yes, a rocket scientist!—tried to show his son why his answer was wrong, but ended up proving that it was right. The father contacted the examination board, which apologized, and the story later made the front page of *The New York Times*. You might be able to guess the correct answer without any help, but if you want a tip, you might find it useful to place two square pyramids together, as shown below.

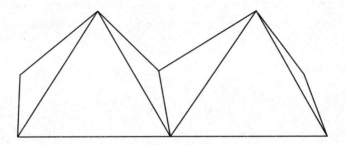

The next question was also set in an exam. In 1995, the inaugural Trends in International Mathematics and Science Study—TIMSS—became the largest international assessment of the abilities of students around the world, testing students across 41 education systems. The following question was given to 18-year-olds in 16 countries studying "advanced" math.

Overall, only 10 percent of students got the right answer. Top of the class was Sweden, where 24 percent got it correct. In the US and France, the score was a measly 4 percent.

I'm including the problem here because it has a "simple" solution, and requires no technical math beyond what should be known by a 14-year-old. Sometimes knowing too much math can be a disadvantage.

72

THE ROD AND THE STRING

A string is wound symmetrically around a circular rod. The string goes exactly four times around the rod. The circumference of the rod is 4 cm and its length is 12 cm. Find the length of the string.

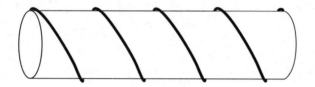

Just as the circle is the simplest two-dimensional shape, the sphere is the simplest three-dimensional object. The Earth is roughly spherical. In fact, for the next problem we can assume that the Earth is a perfect sphere.

A well-known puzzle concerns a person who walks one hundred miles due south, one hundred miles due west, and finally one hundred miles due north, only to return to the very point where they started. The question asks: *What color is the bear?*

White, of course. The only bear species that lives in a region where this three-legged trek is possible is the polar bear. If the traveler starts at the North Pole, a trip one hundred miles due south, one hundred miles due west, and one hundred miles due north traces a triangle that returns them to their point of departure, the North Pole. Which leads us to the following shaggy tale.

WHAT COLOR IS THE BEARD?

A person who walks 10 miles due north, then 10 miles due west, and then 10 miles due south, finds themselves back where they started. They are not at the South Pole.

What color is the beard?

The remaining compass direction is east.

AROUND THE WORLD IN 18 DAYS

In an updated version of Jules Verne's novel *Around the World in Eighty Days*, Phileas Fogg circumnavigates the globe by airplane. His departure point is London, and he travels through the same countries as in the original story: Egypt, India, Hong Kong, Japan, and then across the US. He leaves at noon on October 2nd, and counts 18 days until he gets back. What date does he arrive in London?

To celebrate the end of this chapter, a shot of whiskey.

A WHISKEY PROBLEM

A full whiskey bottle has a height of 27 cm and a diameter of 7 cm, and contains 750 cm³ of whiskey. Like many bottles, it has a dome-like indentation at the bottom.

You have a drink, fall asleep, and wake up to discover that the bottle now contains whiskey to a height of only 14 cm. When you turn the bottle over, the height of whiskey is 19 cm.

How much whiskey is still in the bottle, in cubic centimeters?

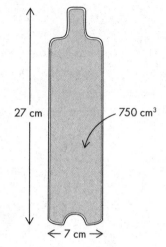

27 cm

750 cm³

← 7 cm →

We know from a previous question that the area of a circle is πr^2, where r is the radius and π is 3.14 to two decimal places. The volume of a cylinder is $\pi r^2 \times h$, where r is the radius and h is the height.

Try to solve the whiskey problem without writing down a single equation. If you're having trouble, you might be tempted to drink some more. Don't pour yourself too much, though!

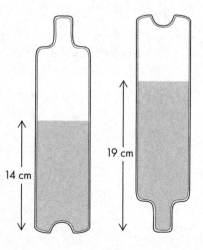

14 cm

19 cm

Tasty teasers

Pencils and utensils

1)

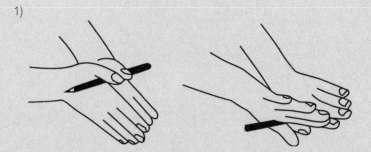

Put your hands together, with a pencil between your thumbs, as shown above left. Can you twist your hands so you get to the position shown above right, in which the pencil is underneath, resting between your palms and thumbs? During the twist you must not let go of the pencil.

2)

Two iron rods are on a table. One is magnetized, with a pole at each end. How can you tell which one is the magnet if all you are allowed to do is move them around the table with your fingers, without using any other instrument or lifting them up?

3)

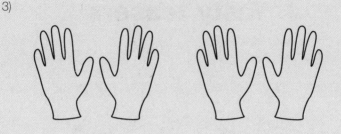

A dentist has three patients to see, but only two pairs of sterile gloves. How can the dentist attend to all three patients (wearing gloves) without the risk of transferring germs from any one person to another person via contaminated gloves? You can assume she has an assistant on hand to help her take the gloves on and off.

4)

There's a classic puzzle that asks you to make a square with two forks on a rectangular table. The solution is to place the two forks perpendicular to each other at one corner of the table; using two sides of the table you create the square. But can you take the same two forks and make four squares instead?

5) How can you cut a hole in a postcard that is big enough for a person to get through?

A wry plod

WORD PROBLEMS

I ♥ math. I also ♥ words. In fact, it is common for math-loving folk to display a partiality to words, puns, and wordplay. Math is all about playing with combinations. Wordplay just swaps abstract symbols for symbols in word form. This part of my book is about having fun with our ABCs. You will confront conundrums about writing constraints, translation, punctuation, fonts, and anagrams. (Look up! Can you crack "A wry plod"?) My initial task for you, though, right now, is to find what is so linguistically atypical about this paragraph. An unusual thing is going on, without a doubt. If you look particularly scrupulously at what is in front of you, it will dawn on you fairly soon. A tip: Go through this paragraph word by word. Do it now, prior to carrying on.

I hope you solved it with ease. I mean, without *E*'s. The previous paragraph did not contain the most common letter in the English language, the alphabetical character known to its detractors as the *filthy fifth glyph*.

A piece of text that omits a particular letter, or set of letters, is called a *lipogram* and is an example of "constrained writing," a technique in which writers enforce some kind of pattern on their writing, or follow a rule that forbids certain things. The earliest known lipogram is a Greek poem from the sixth century BCE, written deliberately without the letter sigma. The genre's most impressive achievement, however, is the French writer Georges Perec's novel *La Disparition*, from 1969. This 300-page book contains no occurrences of the letter *E*, which is even more common in French than it is in English. A good lipogram disguises its constraint. A sign of Perec's virtuosity in composing fluent *E*-less text is that when *La Disparition* came out, some readers did not notice that the letter was missing, even though its absence was a central theme in the plot. (The literal translation of *La Disparition* is *The Disappearance*, but since these words include three *E*'s, Gilbert Adair titled his 1994 English translation of the book, *A Void*.)

Many writers are drawn to literary constraints. Abiding by a strict rule can be unexpectedly liberating, and produce unusual, ingenious, and beautiful results. Indeed, Perec was a member of a collective of writers and mathematicians, Oulipo, the Ouvroir de Littérature Potentielle, or the Workshop of Potential Literature, which over the past 50 years has created many literary works based on mathematical rules. In the spirit of Oulipo, the puzzles in this chapter also celebrate the playful interplay between mathematics and language. Word up.

The inverse challenge to writing text without an *E* is to write text in which the only vowel allowed is an *E*. Such as *Persevere ye perfect men, Ever keep these precepts ten*, instructions for abiding by the Ten Commandments.

THE SACRED VOWELS

Below are five sentences with the vowels and spaces taken out.
Your task is to reinsert the vowels to recreate the sentences. Each
sentence uses one vowel only. The five vowels—A, E, I, O, and U—
each have their own sentence.

[1] L L N T H J Z B L C H W S S W T P R S R V S T H N B L C H
S B R

[2] R T H D X M N K T T W L F S D W N T W B W L S F P R K W
N T N

[3] N R B L S S G N W S T L M B S H N K N D H S J M T R T S
S S N C K

[4] B D S D R N K C H M G L G S R M P N C H, P C H C K S H
M M S B R N C H, S L M P S

[5] T H S R C H D S H S F G S N C N G, W H C H F N S H W
H L S T S W G G N G G N

Each sentence involves food and drink.

If you find univocalic text enthralling, as I do, I strongly recommend
Eunoia, from 2001, by the Canadian writer Christian Bök. A masterpiece
of constrained writing, it has five chapters, each of which allows only a
single vowel. The previous problem was inspired by that book.

In fact, as a subsidiary challenge, choose a vowel and try to write a
meaningful sentence of at least seven words that uses only that vowel. You

will find *A* and *E* sentences the easiest, *I* and *O* more difficult, and *U* the hardest. U have been warned.

Mary Youngquist was the first woman to get a PhD in organic chemistry from MIT. A fan of puzzles, she was editor of the US National Puzzler's League newsletter for six years in the 1970s. She also wrote the following poem, which hides a very simple constraint.

(77)

WINTER REIGNS

Shimmering, gleaming, glistening glow
Winter reigns, splendiferous snow!
Won't this sight, this stainless scene,
Endlessly yield days supreme?
Eying ground, deep piled, delights
Skiers scaling garish heights.
Still like eagles soaring, glide
Eager racers; show-offs slide.
Ecstatic children, noses scarved
Dancing gnomes, seem magic carved
Doing graceful leaps. Snowballs,
Swishing globules, sail low walls.
Surely year-end's special lure
Eases sorrow we endure,
Every year renews shared dream,
Memories sweet, that timeless stream.

What simple rule governs this verse?

The poem may have little literary merit, a comment I'm sure Youngquist would have agreed with. Its charm and inventiveness lie in the way that, despite such a stringent constraint, it reads like a poem, and makes sense.

Almost all of Oulipo's output is in French. Two modern champions of constrained writing in English are the Americans Mike Keith and Doug Nufer. The former is best known for his work in "Pilish," a rule that states that word lengths must follow the digits in *pi*, the number whose first six decimal digits are 3.14159 and whose remaining digits continue ad infinitum without ever repeating a pattern. In a piece of Pilish text, the first word has 3 letters, the second 1 letter, the third 4, and so on. Keith's book, *Not A Wake*, follows the first 10,000 digits of *pi*. It begins: *Now I fall, a tired suburbian . . .*

One mathematical pattern evident in almost all text—books, articles, emails—is that about half the words used appear more than once. Doug Nufer's magnum opus is a 40,000-word novel, *Never Again*, in which every word appears exactly once. (He counts plurals as different words.) To break the universal law of word frequency while still saying something meaningful at such length is a remarkable achievement. His book starts: *When the racetrack closed forever I had to get a job.* And ends: *Worldly bookmaker soulmates rectify unfair circumstance's recurred tragedies, ever-moving, ever-hedging shifty playabilities since chances say someone will be for ever closing racetracks.*

If you try to write a paragraph in Pilish, or one with no repeated words, you will quickly appreciate quite how impressive Keith and Nufer's work is.

FIVE DEFT SENTENCES

Each of the sentences below is written according to a different constraint. Can you work out what each constraint is?

[1] Deft Afghans hijack somnolent understudies.

[2] Dennis, Nell, Edna, Leon, Anita, Rolf, Nora, Alice, Carol, Lora, Cecil, Aaron, Flora, Tina, Noel, and Ellen sinned.

[3] Quiet Pete wrote poor poetry. Had alfalfa dhal (half a flask). Mmmmm.

[4] I do not know where family doctors acquired illegibly perplexing handwriting.

[5] a wise unicorn swims in a ravine, saves an anemic racoon.

THE CONSONANT GARDENER

Plant a consonant in each cell to make five words.

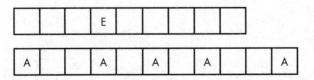

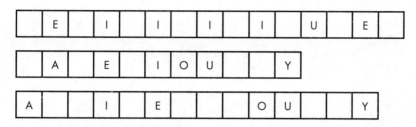

The word at the top of this page is not even the longest word in English that alternates between vowels and consonants. That honor goes to "honorificabilitudinitatibus," which means "honorableness" and appears in Shakespeare's *Love's Labor's Lost*.

A kangaroo carries a joey in its pouch. Likewise, kangaroo words carry smaller versions of themselves.

KANGAROO WORDS

A kangaroo word contains the letters, in the correct order, of a synonym of itself. For example, BLOssOM includes the word "bloom." For each of the kangaroo words below, find the synonym it contains.

Chicken

Contaminate

Deceased

Fabrication

Honorable

Instructor

Separate

Myself

Precipitation

Satisfied

THE TEN-LETTER WORDS

Each of these words has 10 letters. What else do they have in common?

Keennesses
Nagnagging
Rememberer
Rerendered
Sleeveless

TEN NOTABLE NUMBERS

Find the following number words.

[1] The only number word that accurately describes how many letters it has.
[2] The ten-letter number word that contains ten different letters.
[3] The largest number word without an *N*.
[4] The only number word with every letter in alphabetical order.
[5] The only number word with every letter in reverse alphabetical order.
[6] The smallest whole number that includes every vowel—*A, E, I, O, U*—in that order.

[7] The six-letter word that describes where its first letter appears
    in the alphabet.
[8] The lowest whole number to include an *A*.
[9] The lowest whole number to include a *B*.
[10] The lowest whole number to include a *C*.

In 1982, the British electronics engineer Lee Sallows devised a mind-boggling new literary constraint: the self-enumerating sentence, in which the sentence accurately lists what it contains. Here is his first attempt:

Only the fool would take trouble to verify that his sentence was composed of ten A's, three B's, four C's, four D's, forty-six E's, sixteen F's, four G's, thirteen H's, fifteen I's, two K's, nine L's, four M's, twenty-five N's, twenty-four O's, five P's, sixteen R's, forty-one S's, thirty-seven T's, ten U's, eight V's, eight W's, four X's, eleven Y's, twenty-seven commas, twenty-three apostrophes, seven hyphens and, last but not least, a single !

In subsequent years Sallows devised many variations, such as this "pangram": a sentence that contains every letter of the alphabet.

This pangram contains two hundred nineteen letters: five A's, one B, two C's, four D's, thirty-one E's, eight F's, three G's, six H's, fourteen I's, one J, one K, two L's, two M's, twenty-six N's, seventeen O's, two P's, one Q, ten R's, twenty-nine S's, twenty-four T's, six U's, five V's, nine W's, four X's, five Y's, and one Z.

A natural conclusion to Sallows's work on self-enumerating sentences is the self-enumerating question.

THE QUESTIONS THAT COUNT THEMSELVES

[1] How many letters would this question contain if the answer wasn't already seventy-one?

[2] How many letters would the next question contain if the answer wasn't already seventy-eight?

[3] How many letters would the previous question contain if the answer wasn't already seventy-six?

[4] How many letters would the next question contain if the answer wasn't already ninety-two?

[5] How many letters would these two questions jointly contain if the answer wasn't already one hundred sixty-five?

The questions sound self-contradictory, like one-liners in a stand-up comedy routine. I'll help you out with the first one. The question reveals that there are 71 letters in the sentence. What Sallows is asking you to do is to substitute "seventy-one" with a number word that also describes the correct number of letters in the new sentence. For example, "seventy-two" can't be the answer, since "seventy-two" has the same number of letters as "seventy-one," meaning that the new sentence continues to have 71 letters, making the sentence incorrect. The number that works is "seventy-three," since "seventy-three" has two more letters than "seventy-one," thus bringing the total number of letters in the sentence to 73.

The Belgian recreational mathematician Eric Angelini created the following puzzle, which also plays with words, numbers, and self-reference.

THE SEQUENCE THAT DESCRIBES ITSELF

Here are the first four terms in a sequence of number words:

SIX, ONE, EIGHT, FIVE . . .

Each of the words in the sequence describes the number of letters you must count until you reach the next E in the sequence.

SIX, ONE, / E / IGHT, FIVE / . . .

   6   1   8

 (SIX)   (ONE) (EIGHT)

It takes 6 letters to get to the first E, it takes another 1 letter to get to the second E, it takes a further 8 letters to get to the third E, and so on. These numbers—6, 1, 8 . . .—in words are precisely the sequence itself.

If each term in the sequence is the lowest possible number of letters to the next E, find the next six terms in the sequence.

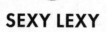

SEXY LEXY

Let a number be known as *lexy* if the number of digits in its decimal expansion is equal to the number of letters it has when it's described in words.

For example, "eleven trillion" is *lexy* because "eleven trillion" has 14 letters and 11,000,000,000,000 has 14 digits.

What is the lowest *lexy* number?

LETTERS IN A BOX

Find the words by placing letters in the cells. Hyphenated words are acceptable.

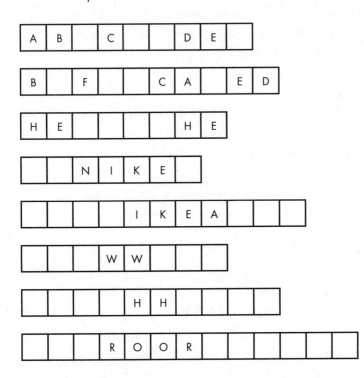

WONDERFUL WORDS

[1] From which word can you take away the whole, and yet have some left?

[2] The word "stressed" is the longest English word to have which property?

[3] What links the words "natal," "nice," and "reading"?

[4] Which common English word becomes plural when an *A* is added to its start?

[5] Which familiar word of five letters becomes shorter if you add two letters to it?

[6] Which English word describes the absence of a person or thing, and yet when you insert a space between two letters of this word, the two resulting words describe that person or thing as being present at this very moment?

[7] Name a five-letter word that is pronounced the same when four of its five letters are removed.

[8] Why does day begin with a *D* and end with an *E*?

[9] In what way do these six adjectives form a sequence? Unpredictable, sexual, direct, warlike, jolly, gloomy

[10] What will always stay the same however many letters you take from it?

Here's a simple sentence: Buffalo!

A context for this sentence might be during a country stroll when you suddenly see a herd of the animals approaching.

Here's another sentence: Buffalo buffalo.

The verb "to buffalo" means to intimidate, or baffle. In the context of your being intimidated, or even baffled, by some buffalo, the sentence clearly makes sense.

Another one: Buffalo buffalo buffalo.

The first word now refers to the city of Buffalo, New York. Again the sentence makes sense. Buffalo from the city of Buffalo intimidate.

In fact, it is possible to make grammatically correct and meaningful sentences that consist solely of the word "buffalo/Buffalo," repeated as many times as you desire. Don't be buffaloed by the buffaloes. I heard a herd ahead.

(88)

LIFE SENTENCES

[1] Explain the meaning of the following sentence.

Buffalo buffalo Buffalo buffalo buffalo buffalo Buffalo buffalo.

[2] Punctuate the following sentence so it makes sense.

John where James had had had had had had had had had had had the teacher's approval.

[3] On what day was the following sentence written?

When the day after tomorrow is yesterday then "today" will be as far from Sunday as that day was which was "today" when the day before yesterday was tomorrow.

The next problem concerns a much simpler sentence.

(89)

IN THE BEGINNING (AND THE MIDDLE AND THE END) WAS THE WORD

Which single word can be placed in each of the eight marked spaces to make eight meaningful sentences?

_ I kicked him in the leg yesterday.
I _ kicked him in the leg yesterday.
I kicked _ him in the leg yesterday.
I kicked him _ in the leg yesterday.
I kicked him in _ the leg yesterday.
I kicked him in the _ leg yesterday.
I kicked him in the leg _ yesterday.
I kicked him in the leg yesterday _.

The following puzzles concern the shapes of letters, words, and sentences. Just my *type* of problem.

LOOKING AT LETTERS

[1] The single *L* and two *Z*'s shown below make up a picture of which letter?

[2] What symbol comes next?

NYXM

[3] Decipher these three common English words.

ZO-ZO
ZOOZ
OON
ZCZ

A MATTER OF REFLECTION

WHEN THE AUTHOR WROTE THIS QUESTION, THEY HID A WORD IN IT THAT HAS A SPECIAL PROPERTY. IT IS THE ONLY WORD IN THE TEXT THAT LOOKS EXACTLY THE SAME WHEN YOU READ IT IN A MIRROR WITH THE PAGE TURNED UPSIDE DOWN. ALL THE OTHER WORDS WILL HAVE AT LEAST ONE LETTER THE WRONG WAY ROUND. CAN YOU DISCOVER THE SPECIAL WORD WITHOUT USING A MIRROR?

THE BLANK COLUMN

When the title of this book, shown below, is typed repeatedly in the same paragraph, a peculiar phenomenon is visible: A blank column (marked in grey) runs through the text.

Perilous problems for puzzle lovers! Perilous problems for puzzle lovers! Perilous problems for puzzle lovers! Perilous problems for puzzle lovers! Perilous problems for puzzle lovers! Perilous problems for puzzle lovers! Perilous problems for puzzle lovers! Perilous problems for puzzle lovers! Perilous problems for puzzle lovers! Perilous problems for puzzle lovers! Perilous problems for puzzle lovers! Perilous problems for puzzle lovers!

Would a blank column appear when any sentence is repeated ad infinitum, or did I deliberately choose the title because it creates this strange effect? (Consider only sentences that are shorter than the length of the line.)

By the time you opened this book, you had already seen a puzzle based on the outline of a letter. The shape emblazoned on the back cover is a capital letter cut out of paper and folded. The challenge is to deduce the original letter, which you are told is not the first one that comes to mind. This puzzle has pedigree. It was the first original problem devised by Scott Kim, then aged 12, who would go on to become one of the most important puzzle designers in the US. One of Kim's trademarks is to take an idea and explore its many variations.

(93)

WELCOME TO THE FOLD

The strange-looking alphabet shown opposite has been created by cutting 26 standard-looking capital letters from a piece of paper and folding each of them once. For example, the A is actually an H.

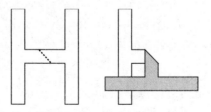

Can you unfold the other 25 letters to discover which letter of the alphabet each one actually is?

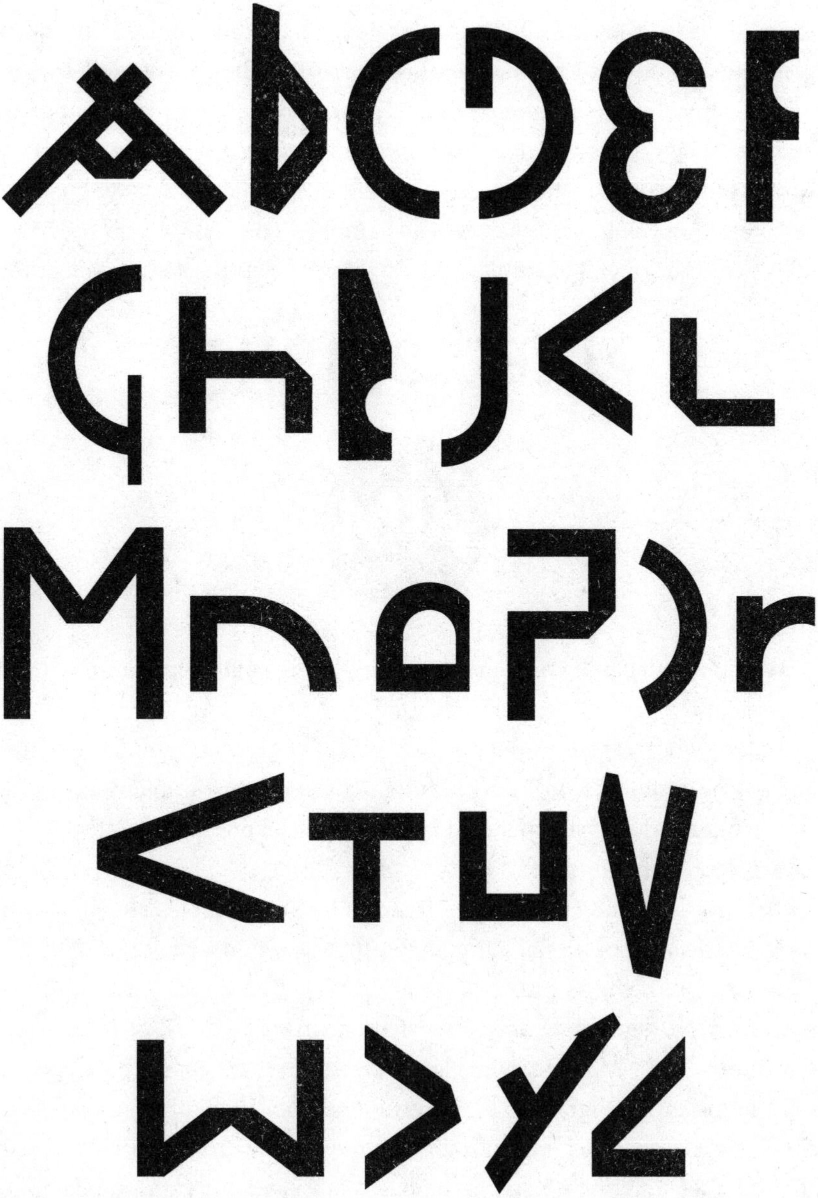

A B C D E F

G H I J K L

M N O P Q R

S T U V

W X Y Z

Kim's interest in letters and typefaces flourished while he was an under-graduate at Stanford University in the mid-1970s, where he studied mathematics, music, and computer science. Fascinated by the intersection of art and technology, and particularly user-interface design, he found an innovative way to link his mathematical and artistic sides: by devising a symmetrical calligraphy in which words can be read in more than one way. His renderings of the words below, for example, read the same if you turn them upside down.

Words which can be read in different ways are called "ambigrams." (Kim is considered a co-inventor of the ambigram, together with the artist John Langdon, who started doing similar work independently at around the same time.) Kim has designed hundreds of ingenious ambigrams, and is such a master that if you give him any word he can pretty much turn it into an ambigram on the spot.

The most common type of ambigram (like the examples shown above) uses 180-degree rotational symmetry—that is, when you rotate it half a turn the word is the same. The capital letters I, N, O, and S have rotational symmetry, so, depending on the typeface, a word like NOON is a natural ambigram. As is SWIMS and, in lower case, *chump*.

Creating an ambigram is an entertaining challenge. It requires you to think about how much visual information you need to suggest a letter of the alphabet. You might need to remove visual elements from some letters, but not so much that the letter is unrecognizable, or you might need to add visual elements, but not so many that the eye is overwhelmed. The form relies on trickery and suggestion. You can contaminate a letter with quite a lot of extraneous detail that the eye will choose not to see if the salient features are there.

The next puzzle is Scott Kim's introduction to how to draw in this style.

MY FIRST AMBIGRAM

Find a way to write the following words so that they read the same upside down:

USA
VISTA
CHILD

These three words present a perfect initiation into the pain and pleasure of ambigram design. They already have some helpful symmetries. Try to come up with several solutions for each word. There is no "correct" answer, but the more you do it, the more elegant your solution will become. You might have to stick with capitals, or use lowercase letters, or use a mixture of the two.

Your next ambigram challenge is to write your name. Lucky for you, Bob, Una, and Wim. Sorry Aloysius, Josephine, and Waldemar.

Scott Kim is most famous for his ambigrams, but he has also designed a huge number and variety of excellent puzzles. Some of my favorites are based on the shape and form of letters. In the next one, he asks us to re-create words with almost no information at all.

BOXED PROVERBS

Can you decipher these 10 familiar sayings? Each letter in the saying has been replaced by a black box the same height and width of that letter.

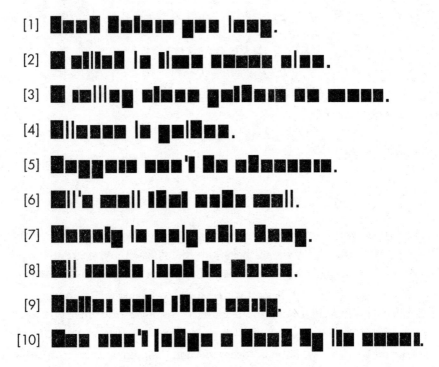

This puzzle is also a kind of optical illusion. If you squint at the page, the phrases become easier to see. Our eyes glide first to the letters that stick up or down. Even though we read from left to right, reading is made easier thanks to the up-and-down rhythm of ascenders and descenders, which is why it is easier to read large amounts of text in lowercase than in uppercase.

Scott Kim's boxed proverbs puzzle is a variation of a similar pattern-recognition problem in which we are given only the first letter of the word.

(96)

NMRCL ABBRVTNS

Identify the words in each of the abbreviated statements below. Each word is denoted by its initial letter (in upper case), and the statements are all true. For example, the answer to "A H. has 5 F." would be "A hand has 5 fingers."

[1]   W. + S. + S. + F. = the 4 S.

[2]   A. + A. + A. + A. + E. + N. A. + S. A. = the 7 C.

[3]   S. + G. + D. + B. + S. + D. + H. = the 7 D.

[4]   The A. has 26 L.

[5]   A G. has 6 S.

[6]   There are 52 C. in a D.

[7]   A S. has 8 L.

[8]   A C. has 64 S.

[9]   A G. C. has 18 H.

[10] A 1 followed by 6 Z. is a M.

The following question about a Russian family also concerns initials.

To answer it, however, you need to know about the Russian system of *patronymics*, in which a man's middle name is his father's name with the suffix *-ovich, -evich,* or *–ich* added. For example, in Russia I would be Alex

Davidovich Bellos, since my father's name is David, and my sons would both have the middle name Alexovich.

(97)

THE NAME OF THE FATHER

Here are names of males in a Russian family. The first initial is the first name, and the second initial is the patronymic. In this family every father has two sons, the patriarch of the family has four grandsons, and his sons have two grandsons each.

A. N. Petrov	K. T. Petrov	N. K. Petrov
B. M. Petrov	M. M. Petrov	N. T. Petrov
G. K. Petrov	M. N. Petrov	T. M. Petrov
K. M. Petrov	N. M. Petrov	

Draw the Petrov family tree.

From Russia, let's take a day trip across the border to Estonia. The capital city is Tallinn, and the national language is Estonian, which is spoken by 1.1 million people and closely related to Finnish.

If you're an Estonian speaker, the next puzzle will be very easy. If you're not, you will learn something curious about how Estonians tell the time.

TELLING THE TIME IN TALLINN

The times on the following clock faces are written in Estonian.

Kell on üks.   Veerand kaks.   Kell on kaks.

Viis minutit kaks läbi.   Pool neli.   Kolmveerand üksteist.

If an Estonian looks at his watch and tells you one of the following, what time is it?

[1] Kakskümmend viis minutit üheksa läbi.

[2] Kolmveerand kaksteist.

[3] Pool kolm.

[4] Veerand neli.

[5] Kolmkümmend viis minutit kuus läbi.

What's the Estonian for these times?

[1] 4:15   [2] 8:45   [3] 11:30   [4] 7:05   [5] 12:30

You will need to know the following Estonian number words:

6 = kuus   7 = seitse   8 = kaheksa   10 = kümme

The Tallinn time teaser is adapted from a question that appeared in the North American Computational Linguistics Olympiad, a wonderful competition in which high school–age students solve linguistics puzzles.

Linguistics puzzles require a mixture of deductive logic, code-breaking skills, and a sense of how languages work. When you tackle them you feel like a detective deciphering an ancient scroll. Often you have to stab in the dark. Whereas a purely mathematical puzzle might introduce you to a new concept, a linguistics puzzle will often reveal fascinating peculiarities of different languages or cultures.

Here's another puzzle adapted from the North American Computational Linguistics Olympiad. It concerns the Waorani, a group of about 4,000 Amerindians who live in the Ecuadorian Amazon. The Waorani language is an *isolate*—that is, it is not known to be related to any other language.

(99)

COUNTING IN THE RAINFOREST

The following equations involve the numbers from 1 to 10 in the Waorani language. Each underlined sequence represents a different number, and the symbols +, ×, and ^{2} are used in their normal senses of addition, multiplication, and squaring.

[1] mẽña mẽña mẽña mẽña + mẽña go mẽña = ãẽmãẽmpoke go aroke × 2

[2] aroke^2 + mẽña^2 = ãẽmãẽmpoke

[3] ãẽmãẽmpoke go aroke^2 = mẽña go mẽña × ãẽmãẽmpoke mẽña go mẽña

[4] mẽña × ãẽmãẽmpoke = tipãẽmpoke

What are the numbers from 1 to 10 in Waorani?

You don't need to cross the world to find an impenetrable language. Just find an organic chemist.

(100)

CHEMISTRY LESSON

Match the chemical formulas to their names.

C_3H_8, C_4H_6, C_3H_4, C_4H_8, C_7H_{14}, C_2H_2

Heptene, Butyne, Propane, Butene, Ethyne, Propyne.

Well, that was a gas!

It's time to end our linguistic excursions and return to more mathematical territory.

Tasty teasers

Bongard bafflers

Mikhail Bongard was a Soviet computer scientist who studied how computers recognize patterns. In the mid-1960s he devised a style of problem in which 12 images are placed together, as below. The six images on the left conform to a pattern, or rule. The six images on the right conform to a different pattern, often the negative of the rule that applies to the images on the left.

The challenge is to discover the rule that the left obeys and the rule that the right obeys.

Here's an easy one to get you started:

The answer is that, on the left, all the shapes are triangles, and, on the right, they are all quadrilaterals.

The rules in the following problems are all very simple, but finding them can be fiendish.

1)

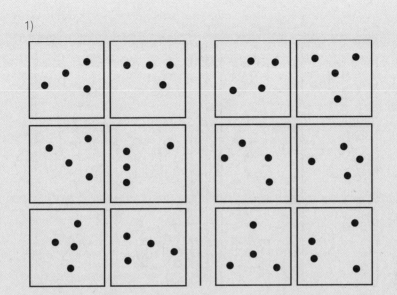

2)

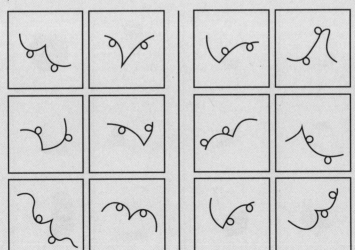

3)

4)

Sleepless nights
and sibling rivalries

PROBABILITY PROBLEMS

The first mathematician to place his hand in the cookie jar was Jacob Bernoulli in 1713.

Okay, so it wasn't a jar. And it wasn't full of cookies.

It was an urn containing 5,000 pebbles, of which 3,000 were white and 2,000 black. (It was a big urn.)

Bernoulli imagined placing his hand in the urn and taking out a pebble at random. He had no way of knowing in advance whether it would be a white one or a black one. Random events are impossible to predict.

But, argued Bernoulli, if he took out a pebble at random, replaced it, took out another pebble, replaced it, took out another pebble, and so on, taking and replacing pebbles for a long enough time, he could guarantee that *on average* he would take out and replace about 3 white pebbles for every 2 black ones. In other words, because the ratio of white to black pebbles in the urn was 3 to 2, in the long run the total number of white pebbles taken out compared to the total number of black pebbles taken out would be approximately 3 to 2.

Bernoulli's insight—that even though the outcome of a single random event is impossible to predict, the average outcome of that same event performed *again and again and again* may be extremely predictable (and approximately equal to the underlying probabilities)—is known as the *law of large numbers*. It is one of the foundational ideas of probability theory, the study of randomness—the mathematical field that underpins so much of modern life, from medicine to the financial markets, and from particle physics to weather forecasting.

Bernoulli's pebble-picking thought experiment also created the template for puzzles in which common objects are plucked at random from urn-like receptacles, like cookies from a cookie jar.

BETTER THAN HALF A CHANCE

You are asked to place 100 cookies—50 made with dark chocolate and 50 made with white chocolate—into two identical jars. Once you have completed this task you will be blindfolded, and you will have to open a jar randomly and take a single cookie from it.

When you are blindfolded, you will not be able to tell which jar is which, nor will you be able to tell the difference between cookies by touch or smell.

You *hate* white chocolate. How do you arrange the 100 cookies between the jars to have the best chance of choosing a dark chocolate one?

SINGLE WHITE PEBBLE

A bag contains a single white pebble and many black pebbles. You and a friend will take turns picking pebbles out of the bag, one at a time, choosing pebbles at random but not replacing them. The winner is the person who pulls out the white pebble.

To maximize your chance of winning, do you go first?

The advantage of going first is that you have a chance to win before your friend does. The disadvantage is that if you don't get the white pebble on your first go, you are presenting your friend with a chance to win, and with one less black pebble in the bag your friend now has a better chance than you just had.

Randomness is a hard concept to wrap one's head around. Indeed, probability is the area of basic math most replete with seemingly paradoxical results, which is one reason why it is a rich source of recreational problems. Some of the best-known puzzles in math are probability posers, and they are notorious because their answers are so counterintuitive. In this chapter we will flip coins, roll dice, and obsess over families' children. Your gut answers to many of the problems will invariably be wrong. Embrace the bewilderment.

Now back to extracting items from darkened receptacles.

(103)

THE JOY OF SOCKS

A sock drawer in a darkened room contains an equal number of red and blue socks. You are going to pick socks out at random. The minimum number of socks you need to pick to be sure of getting two of the same color is the same as the minimum number of socks you need to pick to be sure of getting two of different colors. How many socks are in the drawer?

Next up, another sartorial stumper.

LOOSE CHANGE

I have 26 coins in my pocket. If I were to take out 20 coins from the pocket at random, I would have at least one nickel, at least two dimes, and at least five quarters. How much money is in my pocket?

Problems about selecting items often require you to count combinations. For example, how many ways are there of choosing from a group of two objects?

If the objects are A and B, we could choose: {nothing}, {A}, {B}, {A and B}. Total: four ways.

What about three objects?

If the objects are A, B, and C we could choose: {nothing}, {A}, {B}, {C}, {A and B}, {A and C}, {B and C}, {A and B and C}. Total: eight ways.

To cut a long story short: If we have n objects, there are 2^n different ways of choosing from them.

You may find this information helpful.

THE SACK OF POTATOES

A sack contains 11 potatoes with a combined weight of 2 kg. Show that it is possible to take a number of potatoes out of the sack and divide them into two piles whose weights differ by less than 1 g.

(106)

THE BAGS OF CANDIES

> You have 15 plastic bags. How many candies do you need in order to have a different number of candies in each bag? Every bag must have at least one candy.

The benefits of puzzles are many and varied. They can improve your powers of deduction, introduce you to interesting ideas, and give you the pleasure of achievement.

They are also a "helpful ally" in banishing "blasphemous" and "unholy" thoughts when lying awake at night, wrote Lewis Carroll, the God-fearing author of *Alice's Adventures in Wonderland*.

Carroll, the pen name of Charles Dodgson, a math professor at Oxford University, celebrates puzzle solving as a remedy for self-loathing in *Pillow Problems Thought Out During Sleepless Nights*, a book of 72 recreational problems published in 1893. The book's title is literal. Not only does Carroll divulge that he devised almost all the problems while tucked up in bed (think Victorian nightshirt and nightcap), he also specifies on precisely which sleepless night he thought out which problem.

In the dark hours of Thursday, September 8, 1887, blood must have been pumping wildly around his brain, for that night he came up with the following brilliantly confounding teaser.

A STRATEGY FOR THE DISPLACEMENT OF IMPROPER THOUGHTS

A bag contains a single ball, which has a 50/50 chance of being white or black. A white ball is placed inside the bag so that there are now two balls in it. The bag is closed and shaken, and a ball is taken out, which is revealed to be white.

What is the chance that the ball remaining in the bag is also white?

In other words, you put a white ball in the bag and take a white ball out. The common sense, intuitive answer is 50 percent, because nothing seems to have changed between the original state (mystery ball in bag, white ball outside bag) and the final state (mystery ball in bag, white ball outside bag). If the original ball in the bag had a 50 percent chance of being white, then surely the ball left in the bag must also have a 50 percent chance of being white? Not at all, I'm afraid. The answer is not 50 percent.

In the second edition of *Pillow Problems,* Carroll made a retraction about his nocturnal musings. He rephrased the title, replacing "Sleepless Nights" with "Wakeful Hours" to "allay the anxiety of kind friends, who have written to me to express their sympathy in my broken-down state of health, believing that I am a sufferer from chronic 'insomnia,' and that it is as a remedy for that exhausting malady that I have recommended mathematical calculation." He concludes: "I have never suffered from 'insomnia' . . . [mathematical calculation is] a remedy for the *harassing thoughts* that are apt to invade a wholly-unoccupied mind." [His italics.]

At around the same time in Paris, a French mathematician was harassing his own mind with the conceptual difficulties inherent in basic probability. In his classic textbook from 1889, *Calcul des Probabilités*, Joseph Bertrand set the following problem. Like the Lewis Carroll problem, it involves a situation in which an object is randomly selected and its color observed.

(108)

BERTRAND'S BOX PARADOX

In front of you are three identical boxes. One of them contains two black counters, one of them contains two white counters, and one of them contains a black counter and a white counter.

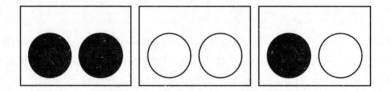

You choose a box at random, open it, and remove a counter at random without looking at the other counter in that box.

The counter you removed is black. What is the chance that the other counter in that box is also black?

If you have never seen this problem before, you will slip on the same banana skin that almost everyone else does.

The most common answer is 50 percent. That is, most people think that if you remove a black counter from a box, half the time the counter left in the same box will be black, and half the time it will be white. The reasoning is as follows: If you choose randomly between the boxes and take out a black counter, you must have chosen from either the first or the third box. If you chose from the first box, the other counter is black, and if you chose from the third box the other counter is white. So the chances of a black counter are 1 in 2, or 50 percent. Your task is to work out why this line of deduction is faulty.

The fact that the answer is *not* 50 percent has led to this problem's being known as Bertrand's box paradox: The correct solution feels wrong, even though it is demonstrably true.

If you are still confused by the last two questions, the following discussion will be useful.

Paradoxes, puzzles, and games have been at the heart of probability theory since its inception. In fact, the first ever mathematical analysis of randomness was written by an Italian professional gambler in the sixteenth century in order to understand the behavior of dice.

Gerolamo Cardano—who also held down the jobs of mathematician, doctor, and astrologer—played a gambling game, *Sors*, in which you throw two dice and add up the numbers on their faces. He saw that there were two ways to throw a 9 (the pairs 6, 3 and 5, 4) and two ways to throw a 10 (the pairs 6, 4 and 5, 5), yet somewhat curiously, 9 was a more frequent throw than 10.

Total is 9

Dice A	Dice B
6	3
3	6
5	4
4	5

Total is 10

Dice A	Dice B
6	4
4	6
5	5

Cardano was the first person to realize that in a proper analysis of the problem, the dice must be considered separately, as illustrated here. When you roll two dice (A and B) there are *four* equally likely ways of getting 9, but only *three* of getting 10. Since there are more equally likely ways to get a 9 than a 10, when you roll two dice again and again, in the long run you will throw a 9 more often than you throw a 10.

The lesson from Cardano that will serve you well in this chapter is: When analyzing a random event *write out a table of all the equally likely outcomes—* also called the "sample space."

Roll with it.

THE DICE MAN DIET

In order to shed some pounds, you implement the following rule: Every day you will roll a die, and only if it rolls a 6 will you allow yourself dessert on that day.

You start on Monday, and roll the die.

On which day of the week are you most likely to eat your first dessert?

DIE! DIE! DIE!

You are given the chance to bet $100 on a number between 1 and 6.

Three dice are rolled. If your number doesn't appear, you lose the stake. If your number appears once you win $100, if it appears twice you win $200, and if it appears on all three dice you win $300. (Like all betting games, if you win you also get your stake back.)

Is this bet in your favor or not?

Try to solve this problem using barely any calculations at all.

A dice throw is a very clear and understandable random event with six equally likely outcomes. Flipping a coin is a very clear and understandable random event with two equally likely outcomes.

(111)

THE PHONY FLIPS

Here are two sequences of 30 coin flips. One I made by flipping an actual coin. The other I made up. Which sequence of heads and tails is most likely to be the one I made up?

[1] T T H T T T T T H H T H H T T H T H H T H H H H T H H H T T

[2] T T H T H H T T T H T H H H H T H H T H H H H T T H T H T T H T

In puzzle-land, giving birth supplants flipping coins as the model for a 50/50 random event. Just as a coin will land heads or tails, a child will be either a boy or a girl. Probability puzzles become much more colorful and evocative when discussing sex balance rather than comparing T's and H's. Indeed, one of the things that makes probability puzzles appealing is that they are usually stated using nontechnical, everyday language.

In the following puzzles we ignore any global variation in sex ratios. Assume that the chance of having a boy or a girl is the same and ignore the possibility of twins.

JUST FOUR KIDS

A couple plans on having four children. Is it more likely they will have two boys and two girls, or three of one sex and one of the other?

⑪⑬

THE BIG FAMILY

The Browns are recently married and are planning a family. They are discussing how many children to have.

Mr. Brown wants to stop as soon as they have two boys in a row, while Mrs. Brown wants to stop as soon as they have a girl followed by a boy.

Which strategy is likely to result in a smaller family? In other words, once they start having children, when they get to the point that one of them wants to stop, is it more likely to be Mr. Brown or Mrs. Brown?

In 2010 I attended a conference of mathematicians, puzzle designers, and magicians in Atlanta. The biennial Gathering 4 Gardner is a celebration of the life and work of Martin Gardner (1914–2010), an American science writer whose most devoted readership comprised members of the above three groups.

One of the speakers, Gary Foshee, took to the stage to give his presentation. It consisted solely of the following words:

I have two children. One of them is a boy born on a Tuesday. What is the probability I have two boys?

Foshee left the stage to silence. The audience was bemused not only by the brevity of the talk but also by the seemingly arbitrary mention of the Tuesday.

What has Tuesday got to do with anything?

Later in the day I tracked Foshee down. He told me the answer: $\frac{13}{27}$, a completely surprising, almost unbelievable result. And, of course, it's all because he mentioned the day of the week.

I wrote about the Tuesday–boy problem later that year in *New Scientist* and on my own blog. Within weeks it had sent the internet into a frenzy of disbelief, indignation, and debate. Why such a strange answer? Why does mentioning Tuesday make a difference? Mathematicians raced to give explanations and refutations. Some agreed with $\frac{13}{27}$, while others argued over semantics or disputed ambiguities in the phrasing. It was my first experience of how fast a good puzzle—or at least a controversial one—can spread around the world.

We'll get to the details of the solution shortly. But first let's look at the Tuesday–boy's antecedents. Conceived by Tom and Michael Starbird, the problem was a new spin on a double-header first posed by Martin Gardner in *Scientific American* in 1959, which itself generated a mailbag of protests.

Mr. Smith has two children. At least one of them is a boy. What is the probability that both children are boys?

Mr. Jones has two children. The older child is a girl. What is the probability that both children are girls?

Read them quickly and you might think they are asking the same question, one about the chance of two sons and the other about the chance of two daughters. Not quite. The phrase "at least one of them" opens up a world of confusion.

Mr. Jones's situation is uncontroversial and simple. If the older child is a girl, the only child of unknown sex is the younger child, who is either a boy or a girl. The probability that Mr. Jones has two girls is 1 in 2, or ½.

Now to the troublesome Smiths. The phrase "at least one of them is a boy" is to be taken mathematically, meaning that either one child is a boy, the other child is a boy, or both children are boys. In this case, there are three equally probable sex assignments of two siblings: boy–girl, girl–boy, and boy–boy. In one out of three cases both children are boys, so the probability is 1 in 3, or ⅓.

The puzzle is tantalizing. The two questions are almost identical, but have drastically different answers, one of which seems to go against common sense. If we know that Mr. Smith has a son, but we don't know whether this son is the older or the younger child, the chance of his having two sons is ⅓. But if we are told that the older (or indeed the younger) child is a son, the chance of his having two sons rises to ½. It seems paradoxical that specifying the birth order makes a difference, because we know with 100 percent certainty that the boy is either the older or the younger.

Scientific American had barely hit the shops when the complaints started to come in. Readers pointed out that the Mr. Smith question, as posed by Gardner, was ambiguous, since the given answer depended on how the information about his children was obtained. Gardner admitted the error and wrote a follow-up column about the perils of ambiguity in probability problems.

To make the Mr. Smith question watertight we need to state that Mr. Smith was chosen at random from the population of all families with exactly two children. If this is the case—and it is arguably how most people understand the question—then the probability that he has two boys is indeed ⅓. But imagine, on the other hand, if Mr. Smith was chosen at random from the population of all two-children families and told to say "at least one is a boy" if he has two boys, "at least one is a girl" if he has two girls, and, if he has one of each, to choose at random whether to say he has at least one boy or at least one girl. In this case, the probability he has two boys is ½. (The chance of having two boys is ¼, the chance of two girls is ¼ and the chance of one of each is ½. So the chance he has two boys is ¼ + [½ × ½] = ½.)

The pitfalls of ambiguity mean that problems like this one must clearly state the process by which the available information is obtained. A further issue is plausibility. What are the real-life scenarios in which a parent of two children would meaningfully say "at least one of my children is a boy" without specifying in any way which child is the boy? These next questions explore this issue.

Tread carefully, boys and girls.

PROBLEMS WITH SIBLINGS

For these questions assume that the parents are chosen at random from the population of families with exactly two children.

[1] Albert has two children. He is asked to complete the following form.

Of the following three choices, circle exactly one that is true for you:
   My older child is a boy
   My younger child is a boy
   Neither child is a boy
   If both are boys, flip a coin to decide which of the first two choices to circle.

Albert circles the first sentence ("My older child is a boy"). What is the probability that both of his children are boys?

[2] A reporter sees Albert's completed form, and writes in the newspaper:

Albert has exactly two children, the oldest of which is a son.

Is a reader of the newspaper correct to deduce that the chance of Albert having two sons is 1 in 2?

[3] Beth has two children.
   You: Can you think of one of your children?
   Beth: OK, I have one in mind.
   You: Is that child a girl?
   Beth: Yes.

What do you estimate are the chances that Beth has two girls?

[4] All you know about Caleb is that he has two children, at least one of whom is a girl. We can imagine asking him whether his older child is a girl, or whether his younger child is a girl, and we know his answer to at least one of these questions will be "yes." Does this insight alter the chances of his having two girls from 1 in 3 to 1 in 2?

[5] A year ago, I learned that Caleb has exactly two children. I asked him one of the following two questions: "Is your older child a girl?" or "Is your younger child a girl?" I know that his answer was "yes," but I can't remember which question I asked! It's a 50/50 chance I asked either question. What are the chances he has two girls?

All these problems were set by the brothers Tom and Michael Starbird.

(Question to Mrs. Starbird: You have two sons. At least one is a boy. What is the probability that both will become mathematicians?)

Tom is a veteran NASA scientist who has worked on many space missions and continues to help operate the Mars Curiosity rover, while Michael is a professor of mathematics at the University of Texas.

Michael also presents home-learning lecture courses. In the early 2000s, when he was preparing one on probability, the brothers came up with their twist on Gardner's two-boy problem (in which Mr. Smith has two children, one of whom is a boy born on a Tuesday). Remarkably, once you mention the day of the week, the probability of Mr. Smith having two boys is no longer ½ or ⅓ but somewhere in between (assuming that Mr. Smith is randomly chosen from the population of all two-children families). The answer seems incomprehensible. How on Earth can knowing the day of the week make a difference to the probability that both children are boys, since the boy under discussion is equally likely to have been born on a Tuesday as he is likely to have been born on any other day?

The Starbirds did not realize quite how their Tuesday–boy problem would leave a trail of misery and irritation. "Many people resist the idea that apparent irrelevancies can have an impact on probabilities," says Michael. "People are so upset by it. Of course, surprises and counterintuitive results are good, so I like people to be startled, but in this case the 'aha' moment sometimes never comes."

Tom says that two of his colleagues at the Jet Propulsion Laboratory reacted to the puzzle with "close to genuine anger. There is something about the puzzle that is seriously disconcerting to some people." He wonders if people get emotionally involved in this type of puzzle because it challenges their basic mathematical confidence—"There may be a subconscious aspect that is upsetting, along the lines of: If my intuition is so wrong for such an easy-to-state problem, am I making errors all the time in other areas?"

The solution to (and further discussion of) the Tuesday–boy problem is provided in the answer to the next question.

THE GIRL BORN IN AN EVEN YEAR

Doris has two children, at least one of whom is a girl born in an even-numbered year. What is the probability that both of her children are girls?

Assume that Doris is chosen at random from the population of two-child families and that babies are equally likely to be born in odd- or even-numbered years.

Here's another puzzle about siblings that tricks our intuition. This time—you may be pleased to hear—their sexes and birthdays are irrelevant.

THE TWYNNE TWINS

A class of 30 students—including the Twynne twins—lines up in single file in the lunch line. The students select their positions randomly; that is, each student has an equal chance of being in any position. The line is single-file, so one of the Twynne twins will be ahead of the other in the line. Call the Twynne twin nearer the front of the line the "first" twin in the line.

If you had to bet that the first Twynne twin will be in a particular position in the line (as in, are they first, second, third . . .) which position would you pick as most likely? In other words, if the 30 students line up in a random order every lunchtime, where in the line would the first twin be most often?

Since the students select their positions randomly, in the long run each student will appear in every position roughly the same number of times. What's counterintuitive, however, is that the Twynne twin nearer the front of the line is more likely to appear in one position than in any other.

The development of probability as a mathematical field led, in the nineteenth century, to the birth of modern statistics: the collection, analysis, and interpretation of data. One important idea from statistics is the "average" in a set of data. You may remember from school that the most common types of average are the *mean* (add up all the values and divide them by the number of terms), the *median* (the middle value if they are all listed in numerical order), and the *mode* (the most common value). Another familiar term when looking at data is the *range*, which is the difference between the highest and lowest values.

A SHOT OF MMMR

Find two sets of five numbers for which the mean, median, mode, and range are all 5. Use only whole numbers.

Averages often mislead. It all depends how you aggregate the data.

LIES AND STATISTICS

A principal is in charge of an elementary school and a high school. Across all pupils the median pupil, ranked by grade, gets a C in math.

The principal introduces a new math syllabus. The following year the median pupil's score has reduced to grade D.

Can you devise a scenario in which the principal has in fact improved everyone's grades?

The principal has all his pupils' scores. Often, however, statisticians do not have all the data and must make estimates based on probabilities.

THE LONELINESS OF THE LONG-DISTANCE RUNNER

You know that there is a race on in your neighborhood. You look out of your window and a runner passes by with the number 251 on their chest.

Given that the race organizers number the runners consecutively from 1, and that this runner is the only one you have seen, what is your best guess for how many runners are in the race? Take "best guess" to mean the estimate that makes your observation the most likely.

Continuing our sporting interlude . . .

(120)

THE FIGHT CLUB

In order to become accepted into the prestigious Golden Cage Club, you need to win two cage fights in a row. You will fight three bouts in total. Your adversaries will be the fearsome Beast and the considerably less fearsome Mouse. In fact, you estimate that you only have a $^2/_5$ chance of beating Beast, but a $^9/_{10}$ chance of beating Mouse.

You can choose the order of your bouts from these two options:

Mouse, Beast, Mouse

Beast, Mouse, Beast

In order to have the greatest chance of winning entry into the Golden Cage Club, which option should you choose?

TYING THE GRASS AND TYING THE KNOT

In the rural Soviet Union, girls used to play the following game. One girl would take six blades of grass and put them in her hand like so.

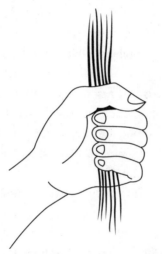

Another girl would tie the six top ends into three pairs, and then tie the six bottom ends into three pairs. When the fist was unclasped, if it revealed that the grass was joined in one big loop, it was said that the girl would marry within a year.

In this game, is it more likely than not that the girl will be told that she will marry within a year?

The search for love involves risk, luck, and probabilities. The following puzzle emerged as a problem about numbers on paper slips, but it would later became a problem about rings on fingers.

THE THREE SLIPS OF PAPER

Three numbers are chosen at random and written on three slips of paper. The numbers can be anything, including fractions and negative numbers. The slips are placed facedown on a table. The objective of the game is to identify the slip with the highest number.

Your first move is to select a slip and turn it over. Once you see the number, you can nominate this slip as having the highest number. If you do, the game is over. However, you also have the opportunity to discard the first slip and turn over one of the remaining slips. As before, once you see the number on that slip, you can nominate it as being the highest. Or you can discard it, which means that you must choose the final slip, and nominate that as having the highest number.

The odds that the first slip you turn over has the highest number is 1 in 3, because you are randomly picking one option in three. But what strategy for the remaining slips will improve your odds of guessing the one with the highest number from 1 in 3?

A follow-up question: There are only *two* slips of paper. Each one has a number on it that was chosen at random. The slips are facedown on a table. You are allowed to turn over only one of them, before stating your guess about which slip has the highest number.

Can you improve your odds of naming the slip with the highest number from 1 in 2?

The answer to both these questions is yes, you can improve your odds with the right strategy. In the case of the problem with two slips only, it's a stunning result, one of the most remarkable in this book. The values on the two slips can be any numbers at all. You only see one of them. Yet it is possible to guess which of the two is the biggest and have a greater than 50 percent chance of being correct. Wow.

But what has all this got to do with marriage?

Like the Tuesday–boy problem, the roots of the slips of paper puzzle can be traced back to Martin Gardner and his *Scientific American* column, which ran monthly from 1959 to 1980. In 1960, he described a variation of the paper-slip puzzle that allowed for as many slips as you like. The table now contains a multitude of facedown slips, each with a different random number on it. Again the aim is to name the highest number. You are allowed to turn over any one slip, but as soon as you do you must either choose it (and forgo seeing the others) or discard it. (The solution for that particular version is also in the back.)

Gardner suggested that this game models a woman's search for a husband. Imagine a woman decides she will date, say, 10 men, each of different suitability, in a year, and imagine that she will indicate after each date whether he should propose. (One assumes that if she gives the okay he will propose, and she will say yes.) The men are the paper slips and the number is their ranking in terms of suitability. Clearly, the woman wants to marry the husband who ranks highest (i.e., she wants to choose the highest number). Once she has started dating, the following rules apply: If she allows someone to propose, she is not allowed to go on a date with the remaining men. And if she rejects a guy, then she cannot return to him later. This behavior models having to either choose a slip and forgo seeing the others, or reject it and move on. Whether or not one agrees with Gardner's simplifications—and even he took them with a pinch of salt—the puzzle has become known as the "marriage problem." The field of math involved here is called "optimal stopping": In a process that could conceivably go on and on, based on what has happened in the past, when is it in your best interests to stop?

The following teaser also first appeared in Martin Gardner's *Scientific American* column, in 1959. It was to become the most notorious probability puzzle of the last 50 years.

(123)

THE THREE PRISONERS

Three prisoners, A, B, and C, are to be executed next week. In a moment of benevolence, the prison warden picks one at random to be pardoned. The warden visits the prisoners individually to tell them of her decision, but adds that she will not say which prisoner will be pardoned until the day of execution.

Prisoner A, who considers himself the smartest of the bunch, asks the warden the following question: "Since you won't tell me who will be pardoned, could you instead tell me the name of a prisoner who will be executed? If B will be pardoned, give me C's name. If C will be pardoned, give me B's name. And if I am to be pardoned, flip a coin to decide whether to tell me B or C."

The warden thinks about A's request overnight. She reasons that telling A the name of a prisoner who will be executed is not giving away the name of the prisoner who will be pardoned, so she decides to comply with his demand. The following day she goes into A's cell and tells him that B will be executed.

A is overjoyed. If B is definitely going to be executed, then either C or A will be pardoned, meaning that A's chances of being pardoned have just increased from 1 in 3 to 1 in 2.

A tells C what happened. C is also overjoyed, since he too reasons that his chances of being pardoned have risen from 1 in 3 to 1 in 2.

Did both prisoners reason correctly? And if not, what are their odds of being pardoned?

This puzzle may feel familiar. At its core is Bertrand's box paradox, which we discussed only a few pages ago.

It is also equivalent to a much more famous puzzle involving two horned, bearded ruminants, a car, and a TV host.

The host is Monty Hall, who presented the game show *Let's Make a Deal* between 1963 and 1990. In the final segment of the show a prize was hidden behind one of three doors and the contestant had to choose the correct door to win it. Here's the "Monty Hall problem" in full.

You have made the final round on Let's Make a Deal. *In front of you are three doors. Behind one door is an expensive car; behind the other two doors are goats. Your aim is to choose the door with the car behind it.*

The host, Monty Hall, says that once you have made your choice he will open one of the other doors to reveal a goat. (Since he knows where the car is, he can always do this. If the door you choose conceals the car, he will choose randomly between which of the other two doors to open.)

The game begins and you pick door No. 1.

Monty Hall opens door No. 2, to reveal a goat.

He then offers you the option of sticking with door No. 1, or switching to door No. 3.

Is it in your advantage to make the switch?

The question was devised in 1975 by Steve Selvin, a young statistics professor at the University of California, for a lecture course he was giving on basic probability. The first time he asked it, all hell broke loose. "The class was divided into warring camps," he remembers. Do you stick or do you switch?

Perhaps the most common initial response is that it makes no difference to switch. If there are two doors left, and we know that the car has been randomly placed behind one of them, it's intuitive to think that the odds of the car being behind either one are 1 in 2.

Wrong! The correct answer is to switch.

In fact, when you switch, you *double your odds* of getting the car. If you stick, your chances of winning the car are 1 in 3. There were three closed doors and you had a 1 in 3 chance of choosing correctly. Your odds don't change when Monty Hall opens another door, because he was always going to open another door regardless of your choice.

If you switch, however, your chances of winning the car rise to 2 in 3. When you choose your first door, as I said in the previous paragraph, you have a 1 in 3 chance of choosing correctly. In other words, there is a 2 in 3 chance that the car is behind one of the other doors. When Monty Hall reveals that one of the doors you did not choose has a goat behind it, there is now only a single choice remaining for the "any other door" option. Therefore, the probability that this remaining door is the one with the car behind it is 2 in 3.

If you are finding this confusing, or counterintuitive, you are not alone. In 1990, the problem appeared in Marilyn vos Savant's column in the large-circulation US magazine *Parade*. She said switch. Her article provoked a deluge of about 10,000 letters, almost all disagreeing with her. Almost a thousand of these letters came from people with a PhD after their names. *The New York Times* reported on the storm on its front page.

The Monty Hall problem requires careful thought. As does its sister puzzle...

THE MONTY FALL PROBLEM

You have made it to the final round of *Let's Make a Deal*. In front of you are three doors. Behind one door is an expensive car; behind the other two doors are goats. Your aim is to choose the door with the car behind it.

The host, Monty Hall, says that once you have made your choice he will open one of the other doors to reveal a goat. (Since he knows where the car is, he can always do this. If the door you choose conceals the car, he will choose randomly between which of the other two doors to open.)

The game begins and you pick door No. 1.

Monty Hall walks to the doors, stumbles, falls flat on his face and accidentally, at random, opens one of the two doors that you didn't pick. It's door No. 2, which reveals a goat.

Monty Hall dusts himself off and gives you the option of sticking with door No. 1, or switching to door No. 3.

Is it in your advantage to make the switch?

Monty Hall's fame has spawned many similar "stick or switch" problems.

RUSSIAN ROULETTE

You are tied to a chair and your crazed captor decides to make you play Russian roulette. She picks up a revolver, opens the cylinder, and shows you six chambers, all of which are empty.

She then puts two bullets in the revolver, each in a separate chamber. She closes the gun and spins the cylinder so you do not know which chambers hold bullets.

She puts the gun to your head and pulls the trigger. Click. You got lucky.

"I'm going to shoot again," she says. "Would you like me to pull the trigger now, or would you prefer me to spin the cylinder first?"

What's your best course of action if the bullets are in adjacent chambers?

What's your best course of action if the bullets are not in adjacent chambers?

Congratulations, you survived. You didn't blow your brains out, although I hope they have been thoroughly massaged over the course of the book.

This chapter concerned probability, an area that underlies much of modern life—from finance to statistics to the frequencies of buses—and one in which it is easy to fall into traps. The more we are aware of how probability can mislead us, the better the decisions we will be able to make in real life. In fact, my aim in the whole of this book has been to present you with puzzles from which there is something to learn, whether a new concept, a clever strategy, or just a simple surprise. I have wanted to kindle creativity, encourage curiosity, hone powers of logical deduction, and share a sense of fun.

Most problems in life don't come with a set of solutions. Thankfully, mine do. And here they are.

ANSWERS

Tasty teasers

Number conundrums

1)

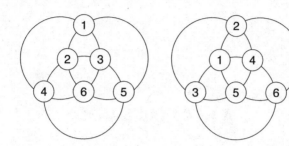

2)

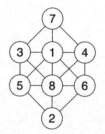

3)

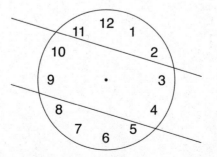

4)

$3 \times 4 = 12$

$13 \times 4 = 52$

$54 \times 3 = 162$

$12 = 3 + 4 + 5$

$2 \times 6 = 3 + 4 + 5$

$3 + 6 = 4.5 \times 2$

$\frac{9}{12} + \frac{5}{34} + \frac{7}{68} = 1$

5)

The first step is to eliminate 5 and 7. If 5 was in the grid, the multiplication(s) including the 5 would be divisible by 5. But the multiplication(s) without the 5 would not be divisible by 5, since no other number between 1 and 9 is divisible by 5. Likewise with 7.

6)

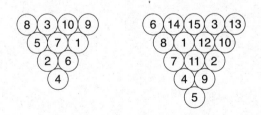

The puzzle zoo

ANIMAL PROBLEMS

① THE THREE RABBITS

② DEAD OR ALIVE

See now the four lines. "Tally ho!"
We've touch'd the dogs, and away they go!

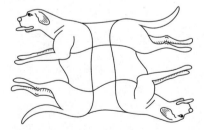

③ GOOD NEIGHBORS

The rabbit was not harmed because it was already dead when the dog picked it up.

④ A FERTILE FAMILY

The single doe has more than 28 trillion descendants, which is about 300 times the estimated total number of humans that have ever lived.

Now to how we get there.

For the first six months of her life, the doe is not fertile. Let $M(n)$ be the total number of does at the end of month n. We have

$M(1) = M(2) = M(3) = M(4) = M(5) = M(6) = 1$

Finally, at the end of the seventh month, she produces six kits, three of which are does, and she produces three more does per month until the end of the year. Note that we can write each monthly total in terms of the total of the previous month.

$M(7) = 1 + 3 = 4$

$M(8) = M(7) + 3 = 7$

$M(9) = M(8) + 3 = 10$

$M(10) = M(9) + 3 = 13$

$M(11) = M(10) + 3 = 16$

$M(12) = M(11) + 3 = 19$

In the second year, she will start to have grandchildren. In the first month of the second year, she adds another three does, and her first litter produces three each. That's four batches of three added to the previous month's total.

$M(13) = M(12) + 4 \times 3 = 31$

In the second month of the second year, she adds another three does, and her first and second litters also produce another three. That's seven batches of three.

$M(14) = M(13) + 7 \times 3 = 52$

$M(15) = M(14) + 10 \times 3 = 82$

$M(16) = M(15) + 13 \times 3 = 121$

You may have spotted a pattern by now—the number of new does is the previous month's total *plus* the total from six months previously, multiplied by 3.

$M(14) = M(13) + 3 \times M(8)$

$M(15) = M(14) + 3 \times M(9)$

$M(16) = M(15) + 3 \times M(10)$

Rewritten as a formula we have $M(n) = M(n - 1) + 3M(n - 6)$. The $M(n - 1)$ part is the running total from the month before, and the $3M(n - 6)$ gives us three new does for every doe that was alive six months ago, which is necessary and sufficient for it to be fertile now. Since the life span of a rabbit is seven years, or 84 months, we want to know M(84).

Calculating this by hand will take too long. Put it in a computer and out will pop 14,340,818,086,651.

But this is not the answer, yet. We need to subtract 1 (the rabbit whose descendants we are trying to calculate) and multiply by 2, since for every doe born there is also a buck.

The final answer is 28,681,636,173,300 descendants.

A bunny bonanza!

⑤ A BUNCH OF HOPS

The rabbits were a clue. The answer is 55, the ninth term in the Fibonacci sequence.

Like many questions of this type, we embark on our solution by simplifying the problem and looking for a pattern. With only two lilies, there is only one way to get from the first to the last lily: a single jump of 1. With three lilies there are two ways: either two single jumps or a double jump; and with four there are three ways: three singles, a single followed by a double, or a double followed by a single.

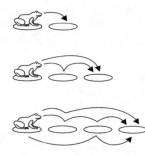

Now consider five lilies, as shown opposite. The frog's first jump will be to either A or B. The total number of ways to get to the last lily is therefore the number of ways from A plus the number of ways from B, which is 2 + 3 = 5.

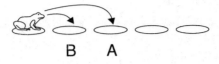

B   **A**

Continuing this line of thought, the number of ways to jump across six lilies is the number of ways to jump across four of them plus the number of ways to jump across five, which is 3 + 5 = 8. The answer for seven lilies is 5 + 8 = 13, for eight the answer is 8 + 13 = 21, for nine it's 13 + 21 = 34, and for ten it's 21 + 34 = 55.

This recursive process—starting with 1, 2, and 3, and generating the next term in a sequence by summing the last two terms—produces the Fibonacci sequence.

⑥ CROSSING THE DESERT

The four Bedouin tribesmen—let's call them A, B, C, and D—depart together, each loaded with five days' worth of water.

At the end of the first day, they each have four days of water left. A gives a day's worth of water to each of his three colleagues, leaving him with a single day's worth of water.

On the second day A returns home (since he is a day away from there, and he has a day's worth of water), while B, C, and D carry on. At the end of the second day, the three of them are now two days out, each with four days of water left. B gives a day's worth of water to the other two, leaving him with two days' worth of water, and the others at capacity with five days' worth.

On the third day B now returns home (which he is able to do since he has two days' worth of water, and the start is two days away) and C and D carry on, reaching a point three days away from the start, with both carrying four days' worth of water. C now gives D a day's worth of water.

On the fourth day, C returns home (which he is able to do since he has three days' worth of water and home is three days away), while D finally reaches the camp, where he delivers the package. D has four days' worth of water left, which is just what he needs to travel back home.

⑦ SAVE THE ANTELOPE

You set off with a full load of five gallons (that's a full tank and four full canisters) and use a gallon to drive 100 miles. Call this point A. Fill the tank, leave the remaining three canisters by the side of the road, and return to base.

Set off again with five gallons and use a gallon to drive the 100 miles to A. Pick up a canister, fill the tank, and with a full load drive another 100 miles. Call this point B. Leave two canisters by the side of the road here. You have two canisters left, and this is enough to get you the 200 miles back to base.

So far you have taken 10 gallons from base, and have two canisters at A and two canisters at B. The gasoline in these four canisters will let you drive for 400 miles, meaning that you only need another four gallons to do the round trip.

Set off for the final time with four gallons: one in the tank and three in canisters. Here is one way you can now split the journey: At A, fill the tank with one canister and take on the other. You now have a full load which will get you the 300 miles to the antelope and a further 200 miles back to B. Once you are back at B, take the two remaining canisters, which is just enough for the 200 miles home.

⑧ THE THIRTEEN CAMELS

The 17-camel puzzle is a lesson about how easily numbers can be used to mislead. The apparent contradiction comes from assuming that the fractions demanded by the father—½, ⅓, and ⅑—add up to 1. But they don't. Written as fractions with lowest common denominator 18 they are ⁹⁄₁₈, ⁶⁄₁₈, and ²⁄₁₈, which add up to ¹⁷⁄₁₈. When the children are given 9, 6, and 2 camels they are *not* being given a half, a third, and a ninth of the total inheritance, as the puzzle suggests. They each receive slightly more. What they do receive, however, is a number of camels that are in the proportions set out by the father—½:⅓:⅑—which when multiplied by 18 are the same as 9:6:2.

The 13-camel problem is a twist on the 17-camel problem in which the wise woman is a bit more provocative. Rather than loaning them a camel, she temporarily takes one away.

The woman hears the children's gripe and says she can solve it if she can confiscate one of their camels. The children reluctantly agree, leaving them with only 12 camels. She gives 6 to the eldest, which is half the total, and she gives 4 to the middle one, which is a third of the total.

That leaves only 2 camels, which she gives to the youngest child. To these camels she adds the one that she confiscated, meaning that the youngest now has 3 camels, a quarter of 12.

The three children thus receive 6, 4, and 3 camels each.

If you read the question carefully, the man does not say clearly that he wants his children to receive a half, a third, and a quarter of the total number of camels. A fair interpretation is that he wants the children to have camels in the proportions $\frac{1}{2}$: $\frac{1}{3}$: $\frac{1}{4}$, or $\frac{6}{12}$: $\frac{4}{12}$: $\frac{3}{12}$, which is indeed the solution the wise woman presents.

⑨ CAMEL VS. HORSE

Ada says: "Get on the other person's animal!"

⑩ THE ZIG-ZAGGING FLY

The simple way to solve this problem is not to think about the distance traveled, but about the time spent traveling. If two cyclists 20 miles apart are cycling at 10 miles an hour, they will meet in one hour. The fly, which is flying at 15 miles an hour, will therefore have traveled 15 miles.

⑪ THE ANTS ON A STICK

Carlos, our anty-hero, falls off last after exactly 100 seconds.

The arbitrary-sounding distances 38.5 cm, 65.4 cm, and 90.8 cm were a sign that you need to think laterally about this puzzle. No puzzle that passes as entertainment is going to require algebra or arithmetic with numbers like 38.5 and 65.4. (Of course, it would be possible to work out the answer by calculating the positions, but that would be messy.)

The piece of insight that makes this an elegant puzzle is this: Think about what happens when two ants collide and both walk back the way they came. If you blur your eyes, this is equivalent to those two ants walking past each other. In other words, this puzzle can be treated as if six ants, each on their own track, are walking to the end of the stick. Another way to visualize it is this: Imagine that each ant is carrying a leaf, and that on each collision they exchange leaves. The leaves will be moving at 1 cm per second in a unique direction. The leaf that takes the longest to fall off the edge is the one that starts with Aggie. So, it will take 100 seconds—1 m at 1 cm per second—until the last ant drops.

Now to the identity of the ant in question. Let's continue thinking about the leaves. Four leaves will fall off the right side (because leaves only move in one direction, and four of them are moving to the right). The last leaf to fall off on the right side is the one that started on Aggie, and we can deduce that the ant carrying it at that point must be the ant that started fourth from the right. Stand up Carlos, that's you. It has to be him because ants can't walk past each other, so the order of the ants on the stick cannot change. The fact that we don't know his position to start with is irrelevant.

⑫ THE SNAIL ON THE RUBBER BAND

During the first second, the snail crawls 1 cm.

In other words, the snail travels $\frac{1}{100,000}$ of the length of the 1-km-long rubber band.

The rubber band stretches instantly to 2 km. During the following second, the snail travels another 1 cm, which now represents $\frac{1}{200,000}$ of the length of the rubber band. After the third second the length of the band is 3 km, so the 1 cm traveled by the snail represents $\frac{1}{300,000}$ of the length of the rubber band. And so on.

If we add up all these fractions, we get the following term:

$$\tfrac{1}{100,000} \left(1 + \tfrac{1}{2} + \tfrac{1}{3} + \tfrac{1}{4} + \ldots + \tfrac{1}{n}\right)$$

This number describes the snail's progress after n seconds, expressed as a fraction of the total length of the rubber band.

If you got this far, congratulations. To finish this proof you need some background knowledge. The term in the brackets, $(1 + \frac{1}{2} + \frac{1}{3} + \frac{1}{4} + \ldots + \frac{1}{n})$, is also known as the "harmonic series," and it gets bigger and bigger the more terms you add, eventually exceeding any finite number. (In my discussion of the jeep problem I mentioned the series $1 + \frac{1}{3} + \frac{1}{5} + \ldots$, which also increases beyond any finite number. If this series increases to infinity, so must the harmonic series. Proving the divergence of the harmonic series is not hard, and a proof can be found easily online.)

We know, therefore, that there exists a number N such that the term $(1 + \frac{1}{2} + \frac{1}{3} + \frac{1}{4} + \ldots + \frac{1}{N})$ exceeds 100,000. After N seconds the expression above describing the snail's progress will be more than 1, which means that the snail will have reached the end of the rubber band. It will take a while! N seconds is longer than the age of the universe. And after N stretches, the rubber band will be so long it will not even fit in the universe.

⑬ ANIMALS THAT TURN HEADS

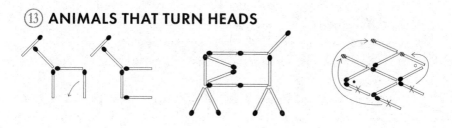

⑭ BANISHING BUGS FROM THE BED

The ceiling gutters need to be positioned beyond the footprint of the bed, and the other way around.

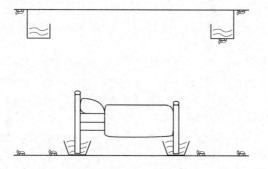

⑮ THE DUMB PARROT

If we are told that the owner of the pet shop never lies, the problem must lie with either the customer or the bird. Here are four possible answers:

The bird is deaf. (She repeats every word she hears, but hears nothing.)

The bird will repeat every word she hears exactly a year after hearing it.

The bird is so intelligent she decided to ignore the owner because she deemed him too stupid.

The customer is lying.

⑯ CHAMELEON CAROUSEL

No, the chameleons will never all have the same color.

Let the number of green, blue, and red chameleons be G, B, and R, and consider what happens when two chameleons meet. If green meets blue, both will become red. In other words, the total number of green lizards becomes $G - 1$, the total number of blue lizards becomes $B - 1$, and the total number of red lizards becomes $R + 2$.

Now consider what that does to the differences between the numbers of lizards of each color—in other words, what happens to $G - B$, $B - R$, and $R - G$:

$G - B$ goes to $G - 1 - (B - 1) = G - B$

$B - R$ goes to $B - 1 - (R + 2) = B - R - 3$

$R - G$ goes to $R + 2 - (G - 1) = R - G + 3$

So, when green meets blue (and by extension when any two chameleons of different colors meet), either the difference between the two other colors stays the same, or it increases or decreases by 3.

When all the chameleons in the colony are the same color, the difference between the two other colors (that's $G - B$, $B - R$, or $R - G$) will be zero.

Now let's enter the numbers given in the question: 13 green, 15 blue, and 17 red. In other words, the difference between green and blue is 2, between blue and red it's 2, and between red and green it's 4. We know that when chameleons meet, the difference between any two colors either stays the same or changes by 3. We can deduce that it is impossible for the difference between any two colors to be 0, since you cannot get to 0 from either 2 or 4 simply by adding or subtracting 3.

⑰ THE SPIDER AND THE FLY

Imagine the glass is made out of cardboard and unroll it, as shown below. From Heron's theorem, the shortest path from the spider to the fly is the distance from the spider to X, where X is 2 cm vertically above the top edge of the cardboard. This distance is the hypotenuse of a right triangle that has sides 6 cm and 8 cm. Using Pythagoras's theorem for right triangles (which states that the sum of the squares of the sides is equal to the square of the hypotenuse), we deduce that the distance in cm that the spider will travel is $\sqrt{(6^2 + 8^2)} = \sqrt{(36 + 64)} = \sqrt{100} = 10$ cm.

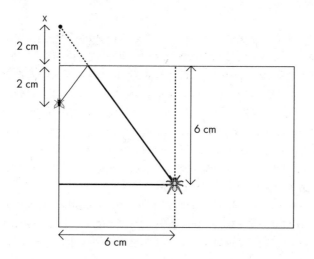

(18) THE MEERKAT IN THE MIRROR

The vertical meerkat is looking at a vertical mirror, as illustrated here.

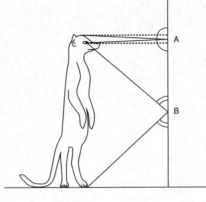

First we're going to work out where the mirror must be hanging, and how long it is, so that the meerkat can see its reflection perfectly framed.

If the meerkat can see the top of its head, light from the top of its head must hit the mirror at a certain point and then rebound into its eyes. Let this point be A. Since the angles of incidence and reflection are equal, the height of A must be halfway between the height of the meerkat's eyes and the height of its head.

Likewise, if the meerkat can see its toes, light from the toes must hit the mirror at a point and then rebound into its eyes. Let this point be B. The height of B must be half the height of the meerkat's eyes from the ground.

If the meerkat's reflection is perfectly framed, meaning that it can see no lower than its toes and no higher than its head, then the mirror must be positioned exactly between A and B. (Which means its length is equal to half the meerkat's height.)

We've worked out the position of the mirror without knowing how far the meerkat is from the wall. Which means that the meerkat can be any distance from the wall, and it will see its reflection perfectly framed. When the meerkat steps back, points A and B stay where they are, meaning that the animal's reflection neither magnifies nor shrinks with respect to the mirror: It stays perfectly framed. Because the angles of incidence and reflection of light are always equal, the angle from A to the meerkat's eyes is always equal to the angle from A to the top of its head, even though the size of this angle will increase as the animal steps backward. Likewise, the angle from B to the meerkat's eyes is always equal to the angle from B to its toes, whatever the size of that angle.

⑲ CATCH THE CAT

Ten days.

Let's begin with an explanation of the solution when there are only four doors. In the grid below, each row displays the possible positions of the cat on each day. The X marks the door I will open on each day.

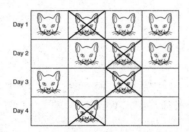

On Day 1 the cat could be behind any door, so there are cats in every cell. I open the second door along. If the cat is there, game over. But if the cat is not there, I can eliminate the chance that the cat will be behind the first door on Day 2, since the only way the cat could be there is if it was behind the second door on Day 1. On Day 2, therefore, the cat can only be behind three possible doors.

On Day 2 I'll open the third door along. If the cat is there, game over. If not, I can eliminate the chance that the cat will be behind the fourth door along on Day 3. I can also eliminate the chance that the cat will be behind the second door on Day 3, since to get there it would have had to have moved from either door 1 or 3, both of which we know do not conceal a cat. We have reduced the choices to two. On Day 3 I open the third door. If the cat is there, game over. If the cat is not there, it must have been behind door 1, which means that by opening the second door on Day 4 I can guarantee that the cat is caught.

Once you work out the four-door solution, it's not so tricky to add an extra door: You can bag the puss in six days by opening the doors 2, 3, 4, 4, 3, 2 in that order.

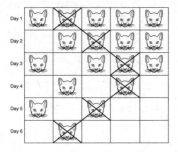

A pattern is developing: Start at the second door along, then open the next door to the right on Day 2, and so on, until you get to the penultimate door. Then return. When there are seven doors, try them in this order: 2, 3, 4, 5, 6, 6, 5, 4, 3, 2. This will ensnare the furtive furry one.

If the number of doors is n, you can always catch the cat in $(n - 2) \times 2$ days.

⑳ MAN SPITES DOG

The mailman would do well to noisily circumnavigate the house, since if the dog is always straining at the leash to be as near to him as possible, it will end up winding its leash around the tree, thus reducing the effective length of the leash until the path is safe.

㉑ THE GERM JAR

Thirty minutes.

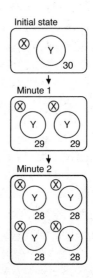

The initial state has a single X and 30 Ys. After one minute, an X will have eaten a Y and doubled, while all the remaining Ys will have doubled. In other words, the population of X is 2 while the population of Y is 2×29. A useful way to visualize this, illustrated right, splits the population into two: Each half has a single X and 29 Ys. A minute later the population can be split in four batches, each with a single X and 28 Ys, and a minute after than into eight (or 2^3) batches, each with a single X and 27 Ys. At the end of the twenty-ninth minute there will be 2^{29} batches, each containing a single X and a single Y. After 30 minutes there will be no Ys left at all.

㉒ THE FOX AND THE DUCK

The fox's best strategy will always be to position itself on the shore as close to the duck as possible. Conversely, the duck's best strategy is to position itself as far as possible from the fox.

We saw that if the duck heads in a straight line for the shore, the fox will catch it. So that plan is out.

But consider what happens when the duck travels in concentric circles around the center of the lake. The fox will also travel in a circle around the lake, trying at all times to be at the closest point to the duck.

Whether or not the fox can keep up with the duck, however, depends on the size of the circle traveled by the duck. If the circular path is close to the edge of the lake, the fox will have no trouble marking the duck. However, if the circle is a small one at the center of the lake, the duck will be able to go one full rotation faster than the fox can.

In fact, since the duck paddles at a quarter of the speed of the fox, both animals will take the same time to make a full rotation when the duck travels a quarter of the distance the fox does. This happens when the radius of the duck's circle is $\frac{r}{4}$, a quarter of the radius of the lake. If the duck paddles in a concentric circle with a radius of less than $\frac{r}{4}$, the fox will not be able to keep up with it and will slowly lag behind.

Imagine now that the duck is traveling in a concentric circle with a radius just under $\frac{r}{4}$ (i.e. just under 0.25r). The fox will lag behind, and after a certain time the duck will be at the position on that circle that is opposite the position of the fox. At that moment the duck will be at a distance of just under 1.25r from the fox, and just over 0.75r from the opposite shore. If the duck makes a direct line to the shore now, it will have to travel this "just over 0.75r" in the same time as the fox runs πr, or 3.14r. Since the fox is four times faster, the duck will make it if "just over 0.75r" is less than $\frac{3.14}{4} r$, which is about 0.78r. Which is possible. Let's say the duck is traveling in a circle of, say, 0.24r. It will always be 0.76r from the shore, so if it starts for the shore when it's opposite the position of the fox, it will reach land before the fox can catch it.

(23) THE LOGICAL LIONS

If there are 10 lions, the sheep survives.

But if there are 11 of them, the sheep will die.

We solve this problem by seeing what happens when the pen has a single lion in it, and then increasing the number by one each time.

One lion, one sheep:

With a single lion in the pen, the sheep doesn't stand a chance. The lion will gobble it up. Result: The sheep dies.

Two lions, one sheep:

Neither of the lions will risk eating the sheep, since if one of them did, it would get drowsy and then be eaten by the other lion. Result: The sheep lives.

Three lions, one sheep:

One of the three lions will happily eat the sheep, because this lion knows that even though it will get drowsy, neither of the other two lions will dare eat it. If one of the other lions did eat the sheep-eater, that lion would then get drowsy and be eaten by the remaining lion. Result: The sheep dies.

Four lions, one sheep:

If one of the four lions eats the sheep, it will become drowsy and the situation becomes equivalent to the situation with three lions, only with the drowsy lion in place of the sheep. Since three lions/one sheep is a scenario in which the sheep dies, none of our four logical lions will eat the sheep. Result: The sheep lives.

You will hopefully see the pattern now. Whether or not a lion eats the sheep depends on the scenario in which there is one fewer lion. If the sheep in the scenario with one fewer lion lives, then the lion will eat the sheep, but if the sheep in the scenario with one fewer lion does not live, the lion will not eat the sheep.

In other words, the mortality of the sheep flips between alive and dead each time you increase the number of lions by 1. If there is an odd number of lions the sheep will die, and if there is an even number the sheep will live. Since the problem states 10 lions, the sheep will live.

24 TWO PIGS IN A BOX

The smaller, subordinate pig eats better. Strength is sometimes a weakness.

This paradoxical conclusion arises because the small pig quickly realizes that when it does press the lever, all it is doing is feeding the big pig, since the big pig will be waiting by the bowl to eat up all the food. Once the big pig is by the bowl, the small pig will not be able to push it out of the way.

The small pig has no incentive to pull the lever, so it doesn't. That job therefore falls to the big pig. When the big pig pulls the lever, the small pig will be waiting by the bowl and will eat as much as he can before the big pig arrives and pushes him out of the way. Since the big pig will get some food, he does have some incentive to press the lever, even though the small pig will often get most of what's in the bowl.

25 TEN RATS AND ONE THOUSAND BOTTLES

Our method will be to feed each rat a mixture taken from different bottles, and to work out which is the poisoned bottle by noting which rats survive and which rats die by the end of the hour. The rats must be fed these mixtures simultaneously, so that after exactly an hour we can see the death toll and deduce which bottle has the poison.

We have 10 rats. After an hour they will be either alive or dead. There are exactly $2 \times 2 \times 2 \times 2 \times 2 \times 2 \times 2 \times 2 \times 2 \times 2 = 2^{10} = 1,024$ possible combinations of 10 alive or dead rats, since there are 10 rats and every rat can be one of two things, alive or dead. The fact that there are 1,024 possible combinations and only 1,000 bottles means that we have more than enough combinations to specify individual bottles.

The question now becomes how to make each unique combination of live and dead rats refer to a unique bottle. The answer is that we use binary numbers— the number base in which only 0 and 1 are used. In the usual decimal number system, the digit farthest on the right is in the "units" column, and then moving left we have the "tens," the "hundreds," the "thousands," and so on, each one being 10 times the one before. In the binary system, the units column counts 1, but moving leftward the columns count 2, 4, 8, 16, and so on, doubling each time, as shown below.

Decimal	Binary
1	1
2	10
3	11
4	100
. . .	. . .
998	1111100110
999	1111100111
1000	1111101000

Our first job is to order all the bottles in binary notation from 1 to 1111101000. We need them all in ten-digit form, so 1 is 0000000001, and 2 is 0000000010.

We also order the rats: We have a "units" rat, a "twos" rat, a "fours" rat, an "eights" rat, and so on up to a "five hundred and twelves" rat.

Finally, we give the wine to the rats.

The units rat gets a cocktail that contains a sip from every bottle whose binary form has a 1 in the units column.

The twos rat gets a cocktail that contains a sip from every bottle whose binary form has a 1 in the twos column.

The fours rat gets a cocktail that contains a sip from every bottle whose binary form has a 1 in the fours column.

And so on.

If the units rat dies, we know that the binary representation of the poisoned bottle has a 1 in the units column.

If the units rat survives, we know that the binary representation of the poisoned bottle has a 0 in the units column.

If the twos rat dies, we know that the binary representation of the poisoned bottle has a 1 in the twos column.

If the twos rat survives, we know that the binary representation of the poisoned bottle has a 0 in the twos column. And so on.

In other words, in a ten-digit binary number that determines which bottle has the poison, the dead rats are the 1s and the live rats are the 0s. Voilà!

The reason that not all the rats will die is because if they did they would have encoded the binary number 1111111111, which is 1,023 in decimal, and our bottles only go up to a thousand. (In fact, it is possible to ensure that two rats survive. To do this we need to order the bottles in such a way that excludes the ten binary numbers with only one 0, as well as the binary number with all 1s.)

Tasty teasers

Grueling grids

1) 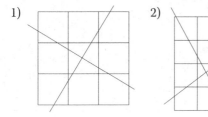 2) 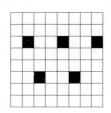 3)

4) Here's one shape that works:

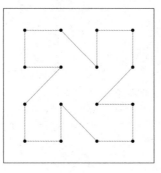

5) There are 20 squares. Orthogonally: nine 1 × 1s, four 2 × 2s (of which only two are shown below; the other two are found by moving these squares into the empty corners), and one 3 × 3. Diagonally, four at a 45 degree angle and two at the lesser angle.

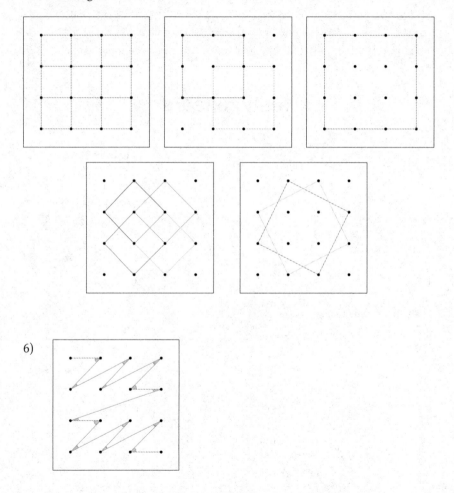

6)

I'm a mathematician, get me out of here

SURVIVAL PROBLEMS

26 FIRE ISLAND

Take a piece of wood from a tree near the leading edge of the fire and light it. Carefully walk to a point on the eastern side of the island—say 100 m from the coastline. Set the forest there alight. Make sure you stay to the west of this new fire, which, aided by the wind, will move eastward at 100 m per hour. The fire will burn out once it reaches the coast, providing you with 100 m of safe space—although be careful not to burn your feet in the ash. The items in your rucksack were red herrings, although the Bible might come in handy as a mat.

27 THE BROKEN STEERING WHEEL

It is possible to turn right by turning left.

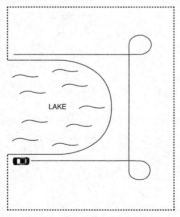

LAKE

(28) WALK THE PLANK

To save the British, $a = 1$, $b = 11$.

To save the French, $a = 9$, $b = 29$.

In Bachet's version of the Josephus problem, with 30 passengers, the mnemonic *Mort, tu ne falliras pas, En me livrant le trépas!* encodes the solution with its vowels. If $a = 1$, $e = 2$, $i = 3$, $o = 4$, and $u = 5$, the order of the vowels reveals the order in which to position the passengers around a circle. Let's call the 15 people you want to save Friends, and the 15 you want to drown Enemies. The first four vowels in the mnemonic are *o, u, e,* and *o* (4, 5, 2, and 4), so you start the ordering around the circle with 4 Friends, then 5 Enemies, then 2 Friends, then 4 Enemies, and so on.

(29) THE THREE BOXES

[1] The key is in the white box.

On the black and red boxes are statements that are opposites, which means that exactly one of them must be true. So the statement on the white box must be false, which means that the white box contains the key.

[2] The key is in the black box.

If the key is in the red box, then all three statements are true. If the key is in the white box, then all three statements are false. So the key must be in the black box, in which case two statements (on the black and white boxes) are true and one (on the red box) is false.

(30) SAFE PASSAGE

Put the ring in the lock box, put a single padlock on it, and send it. Your beloved will receive the box, put a second padlock on it, and return it to you. You will then unlock your padlock and send the box back. Your beloved then opens the second padlock to retrieve the ring.

㉛ CRACK THE CODE

The code is 052. The fourth line eliminates 7, 3, and 8, meaning that the code must be composed from the set of digits 0, 1, 2, 4, 5, 6, and 9. The fifth line reveals that 0 is one of the numbers, and that it must be in either of the first two positions. The third line tells us that 0 must be in the first position, and that one of the other numbers is a 2 or a 6. The first line eliminates 6, since 6 cannot be correctly placed *and* in the first position, because we know that 0 is there. So 2 is the number in the third position. This leaves the middle position, which according to the second line must be 5. We eliminate 4 because if 4 were correct it would be correctly placed.

㉜ GUESS THE PASSWORD

Six guesses is the minimum.

If the digits in the password were independent of each other—that is, if they were allowed to be any digit from 0 to 9, irrespective of the digits in other positions—it would take 10 guesses to open the door because each position could be any of the 10 digits. (You are not going to be able to get a better strategy than typing in 0123456 at your first attempt, 1234567 at your second, 2345678 at your third, and so on until 9012345.) Yet because of the rule that no digit in the password appears twice, the digits *are* dependent on each other, since if a digit appears in one position it cannot appear in another position. The puzzle exploits this dependency.

The best strategy is as follows. Choose any six digits, say, the digits between 0 and 5, and make six guesses that place each of these digits in the same six positions of the password. So, for example:

0123456
1234506
2345016
3450126
4501236
5012346

(I've put 6 as the final digit, but it can be anything you want.)

These guesses will open any password with a 0, 1, 2, 3, 4, or 5 in one of the first six positions. If the password does not have any of these digits in the first six positions, the first six positions must be made up of the other numbers: 6, 7, 8, or 9. However, since there are only four of these numbers, one or two of them would have to appear twice, contradicting the question. In other words, *every* 7-digit password with no repeated digits must have a 0, 1, 2, 3, 4, or 5 in one of the first six positions, and the above strategy works.

㉝ THE SPINNING SWITCHES

At the start, the buttons are in one of three possible combinations: both off, on and off, or off and on.

Our first move is to press both buttons together. If they were both off, now they are both on, and the door opens. If one was on and one was off, the overall picture stays the same. The one that was off is now on, and the one that was on is now off.

The wheel spins. The second move is to press a single button. Either both buttons will now be on, and the door will open, or they will both will be off, and the wheel will spin. When it comes to rest, our third and final move is to press both buttons together, opening the door.

For the harder version, I suggest building yourself a spinning wheel with four switches, and playing around with it. You'll quickly figure out the pattern. In the absence of a spinning wheel, use four coins. Place them on the table in the four compass positions. Heads will represent "on," and tails "off." Pressing a button is now the same as turning a coin over.

Let the following abbreviations stand for the following moves.

E: We turn over *every* coin (or press every button).

O: We turn over any two *opposite* coins, such as, say, north and south. (Or we press two opposite buttons.)

A: We turn over any two *adjacent* coins, say, north and east. (Or we press two adjacent buttons.) Assume that after each move the wheel spins around, unless the door is opened.

Our first seven moves are: EOEAEOE.

If the four coins are all heads in their initial positions (i.e., the buttons are all switched on), the door will open right away. If the coins are all tails, an E will turn them all over, so the buttons will all be on and the door will open. The subsequent moves (OEAEOE) will open the door if exactly two coins are heads at the start. Here's why:

If only two coins are showing heads at the start, either the heads are positioned opposite each other (say, north and south), or they are positioned adjacent to each other (say, north and east). Applying move E (as we did above) does not change the position of the coins—they are still either opposite or adjacent to each other.

If the heads are positioned opposite each other, a move O results in either four heads or four tails. So either the door opens, or we apply E in next move and the door opens. If, on the other hand, the heads had been positioned adjacent to each other, turning over opposite coins leaves a situation in which, again, two heads are adjacent to each other. So, an O followed by an E does not change the situation. We then apply an A, turning over two adjacent coins. Either we now have four heads, in which case the door opens, or no heads, in which case the door will open if we turn all the coins over on the next turn. Alternatively, we are left with a situation in which we have two heads opposite each other. All we need to do again in that case is turn two opposite coins over; if that doesn't work we can turn all the coins over again.

To complete the strategy, we need to find a way of opening the door if there is either one or three heads in the initial placement of the coins. If there was one or three heads at the start, there will still be one or three heads after the moves EOEAEOE have been applied. So for the eighth move, turn a single coin over. Either four coins will now be heads, in which case the door will open, or none of them will be heads, in which case we make our next move an E. Alternatively, exactly two coins will be heads, in which case we repeat the sequence (EOEAEOE) above.

(34) PROTECT THE SAFE

The directors need to put three locks on the safe, and they need two keys for each lock, making a total of six keys.

If the locks are A, B, and C, the keys need to be distributed as follows: One director has the keys for A and B, one those for B and C, and one those for A and C. In this way, no single director can open the door, but any combination of two directors can.

BONUS PROBLEM: THE AVERAGE SALARY

Let the three colleagues be Amy, Ben, and Charlotte. Amy adds a constant to her salary and tells Ben. Ben adds his salary to the total and tells Charlotte. Charlotte adds her salary to the total and tells Amy. Amy now subtracts the constant from the total, and divides by three to get the average, which she tells the others.

(35) THE SECRET NUMBER

The secret number puzzle involves similar thinking to that used in the average salary problem. You ask Lag to think up a number—let's call it L—and he whispers it to you (without the other inmate hearing). You add your gang's number to L, and tell the other inmate the total. The inmate subtracts his number from this total, and whispers it to Lag (so you don't hear). You ask Lag whether or not it's L. If he says yes, you and the inmate are in the same gang. If he says no, you are not in the same gang. You have therefore ascertained whether or not you are in the same gang without either of you having revealed the secret number.

(36) REMOVING THE HANDCUFFS

One person (say, the person with the white string, as shown opposite) needs to thread their string through one of the loops that ties the other person's string to their wrists, and then take it over that person's hand:

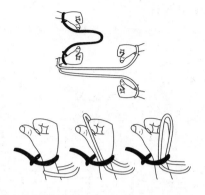

③⑦ THE REVERSIBLE PANTS

First, unpeel your pants so that they're inside out on the rope. (In other words, the pant legs have been reversed once.) Next, grab the hem at the bottom of one pant leg from the inside, pull it through that pant leg and through the other leg. In other words, you are pushing one pant leg through the other. This is a scrunch if you have skinny jeans! (Now the pants have been turned inside out twice, so they are the right way around but with the legs pointing at your feet.) Turn the pants inside out again by pulling each leg through itself (so that the pant legs have been reversed a third time). The pants can now be put on inside out, with the fly at the front.

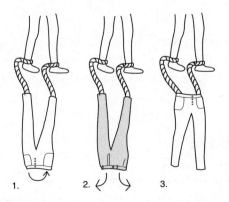

1.   2.   3.

(38) MEGA AREA MAZE

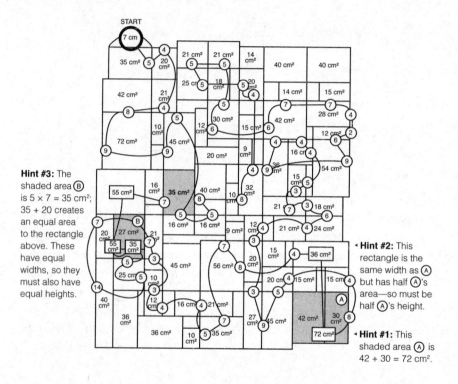

The path to the answer is revealed through the snake of numbers. In the text I explained the first steps: $7 \to 5 \to 4 \to 4$. If that rectangle has area 21 cm² and width 4 cm, then the rectangle to its left with identical height and an area of 42 cm² must have width 8 cm—and so on, until you reach the final area of 35 cm².

(39) ARROW MAZE

Yes, you will get out of the grid.

We solve this problem not by considering this particular grid, but by considering every possible finite size of grid with every possible combination of arrow orientations. Whatever the grid size or arrow positions, you must eventually leave the grid.

Assume, for a moment, that you *don't* get out of the grid. In this case, once you start, you continue moving around the cells forever. But if you are moving, ad infinitum, around a finite number of cells, you must visit at least one of those cells an infinite number of times. Consider this cell. If you visit it an infinite number of times, you must also visit the four cells adjacent to it—that is, above it, below it, and to its left and right—an infinite number of times. Following on from that, you must also visit the cells adjacent to those four, and eventually all the cells in the grid, an infinite number of times.

You cannot visit the bottom right cell an infinite number of times. If the arrow is pointing down, as it is in this particular grid, then it will be pointing to the exit on your third visit and you will be able to leave the grid. If it is pointing in any other direction, you will also leave in at most three visits.

The assumption that you don't get out of the grid must be false. Therefore, you do escape the grid.

(40) THE TWENTY-FOUR GUARDS

Monday

4	1	4
1	PRISON	1
4	1	4

Tuesday

2	5	2
5	PRISON	5
2	5	2

Wednesday

1	7	1
7	PRISON	7
1	7	1

Thursday

	9	
9	PRISON	9
	9	

Friday

5		4
	PRISON	
4		5

㊶ THE TWO ENVELOPES

Did you notice that there is a fire in the room? It pays to be observant in this type of lateral-thinking puzzle. A smart response is to take one of the envelopes and then throw it into the fire. (Making it look accidental, of course, and waiting until it has been fully consumed by the flames.) Tell the king that you would like to choose the one that fell in the fire. In other words, you do not want the one that is still on the table. The remaining one is opened, and it reveals the word DEATH. To keep his honor the king must accept that the other envelope contained the word PARDON.

㊷ THE MISSING NUMBER

You're good at arithmetic, so I expect you to be able to add the numbers from 1 to 100. The fast way to do this calculation is to realize that the sum is the same as $(1 + 100) + (2 + 99) + (3 + 98) + \ldots + (50 + 51)$. In other words, the sum of the first 100 numbers is 101 fifty times, or $101 \times 50 = 5,050$.

If you added up all of the queen's 99 numbers as she read them out, you could easily deduce the missing number, since the missing number would be equal to the difference between the sum you calculated and 5,050.

But adding 99 numbers is quite an effort, and it's easy to make a mistake, even for the arithmetically gifted, especially once you start to hold three- or four-digit numbers in your head.

The clever insight here is to realize that you never need to count above 100. In other words, once you hit 100, rescale back to zero. For example 86 + 15 would be 1, rather than 101. (The mathematical term for this is "counting modulo 100.")

Since you are only keeping a number between 1 and 100 in your head, the adding is not so daunting. Once you get to the end, if your sum (modulo 100) is less than 50, this means the missing number is the difference between your sum and 50. (Had you not counted modulo 100, your sum would be between 5,000 and 5,049, which you would subtract from 5,050 to find the missing number.) If your sum (modulo 100) is more than 50, the missing number is the difference between this number and 150. (Had you not counted modulo 100, your sum would be between 4,950 and 5,000, which you would subtract from 5,050 to find the missing number.)

㊸ THE ONE HUNDRED CHALLENGE

Your strategy is always to say 11 minus the number your cellmate just said. That's why after he started with 8, you replied with 3. Now that he has just said 4, you must reply with 7, which will give a running total of 22. If you proceed in this way, you can guarantee that the running total each round will always be a multiple of 11. Eventually, the running total will be 99, when it will be your cellmate's turn. He will therefore be the first to reach 100.

㊹ THE FORK IN THE ROAD

One solution is to point at a branch of the road and ask the local: "If I were to ask you if this branch leads to the airport, would you say yes?" If it is the route to the airport, both truth-tellers and liars would say "yes," because the liar is forced to lie about her response to the question "does this branch lead to the airport?" If asked the direct question, she would lie and say "no, the branch does not lead to the airport." But the way the question is stated means she must lie about saying no, so she would tell you that she would reply "yes." Likewise, if it was not the route to the airport, both truth-tellers and liars would say "no."

Another solution, which avoids embedding a question within a question, is to point at a road and ask: "Of the two statements 'You are a liar' and 'This branch leads to the airport,' is one and only one of them true?" A truth-teller would say yes if the branch does lead to the airport, and no if not. A liar would also say yes if the branch leads to the airport, since in that case *both* statements would be true, making the correct answer to the question "no." But since she is a liar she must say "yes."

(45) BISH AND BOSH

Look again at the first solution given to the previous problem: You point at a branch of road and ask "If I were to ask you if this route leads to the airport, would you say yes?" Both liars and truth-tellers would say "yes" if the branch did lead to the airport. Now imagine switching yes for no: You point at a branch of road and ask: "If I were to ask you if this branch leads to the airport, would you say no?" In this case, both truth-tellers and liars would say "no" if the branch leads to the airport. In other words, if you ask the question "If I were to ask you if this branch leads to the airport, would you say X?," where X is either yes or no, a response of "X" means that you have pointed at the airport branch. This insight leads to the answer to this problem.

Point at a branch and ask: "If I were to ask you if this route leads to the airport, would you say bish?" If the response is "bish" then the branch leads to the airport, and if it is "bosh," it leads to the beach. Equally, you could ask: "If I were to ask you if this branch leads to the airport, would you say bosh?" A response of "bosh" means you will make your plane.

(46) THE LAST REQUEST

An example of a successful question is:

"Will you answer 'no' and sentence me to death?"

In other words, you are asking whether both of these statements are true:

[1] The executioner will answer "no."

[2] The executioner will sentence you to death.

The executioner cannot answer "yes," since if he did he would be lying by contradicting the statement that he is answering "no."

So the executioner must respond "no." But if the answer is no then it is not true that both statements are true, so one of them must be false. Statement [1] cannot be false since we know it is true—he did answer "no." So statement [2] is false. In other words the executioner will *not* sentence you to death. Your life is spared!

(47) THE RED AND BLUE HATS

Let the prisoners be A and B. The question states that A sees B's hat, and B sees A's hat. If A's guess is always the color of B's hat, and if B's guess is always *not* the color of A's hat, then at least one guess will be correct.

We can see this in the table below, which lists all the possible color combinations of the two hats. In each case, the correct guesses are in bold.

A's hat	B's hat	A's guess	B's guess	Correct guesses
Red	Red	**Red**	Blue	1
Red	Blue	Blue	**Blue**	1
Blue	Red	Red	**Red**	1
Blue	Blue	**Blue**	Red	1

For the three-player version in which at least one person must guess, and all guesses must be correct, the strategy is as follows:

If a player sees hats of two *different* colors, they stay silent.

If a player sees two hats of the *same* color, they guess the opposite color.

The eight possible combinations of hat colors on three players are RRB, RBR, BRR, BBR, BRB, RBB, BBB, RRR. Using the above strategy, the first six combinations will result in two players passing, and a third guessing their hat color correctly. In the final two combinations, all three will guess incorrectly. In other words, in 6 of 8 equally likely arrangements of hat colors, at least one person is guessing and no one guesses wrongly, giving the prisoners a 75 percent chance of survival.

To get a feel for why this works, think about the number of guesses made across all combinations of hat colors. A guess will be made 12 times. Six times the guess will be correct (once each in RRB, RBR, BRR, BBR, BRB, and RBB), and six times the guess will be wrong (three times each in both BBB and RRR). The correct guesses, however, are spread across six combinations while the six incorrect ones are spread across just two combinations. It's like you're dividing up the good stuff across as many boxes as possible, while squeezing the bad stuff into the smallest number of boxes you can. The same approach works with more than three players, with even more impressive results. As mentioned in the main text, if you play this game with 16 prisoners, the odds of survival are more than 90 percent.

(48) THE MAJORITY REPORT

You should adopt this strategy:

[1] At the start of the recital, and whenever you see that the counter is at 0, commit the name you hear to memory and click once upward, so the counter is on 1.

[2] When the counter is on 1 or greater, click upward if the name you hear is the same as the one in your memory, but click downward if the name you hear is not the one in your memory. In both cases keep the same name in your memory.

This strategy guarantees that the name in your memory after the list has been read out is the name that has been said more than half the time.

To get a feel for why the strategy works, let's see what happens if the warden reads out the following list: A B C A B A A B A. She reads nine names in total, made up of three distinct names.

Name read out	Counter	Name in memory
A	1	A
B	0	A
C	1	C
A	0	C
B	1	B
A	0	B
A	1	A
B	0	A
A	1	A

The name in your memory at the end is A, which does indeed appear more than half the time.

(49) THE ROOM WITH THE LAMP

Let's begin with the situation with three prisoners: A, B, and C. We also know that the lamp is off at the start.

The crucial element in the strategy, from which everything else follows, is that one prisoner will have a different role from all the others. Let's call him the Counter, since his job will be to keep track of who has been in the room, and then to announce to the prison warden that everyone has visited it. The gist is that the regular prisoners will turn the lamp on, and that they will be tallied by the Counter, who will turn the lamp off.

Let's say C is the Counter. The rules for him are:

If the lamp is off, do nothing.

If the lamp is on, turn it off.

While the rules for A and B are:

If this is the first time you've seen the lamp switched off, turn it on.

In all other cases, do nothing.

In this scenario, what happens is that the lamp is turned on at some stage by either A or B, after which C switches it off. Eventually either A or B will turn it back on again (if A switched it on first, then B will switch it on next, or vice versa). When C enters the room to see the lamp on for the second time he can be sure that both the other prisoners have been in the room, and he can holler at the top of his voice, "We have all visited the lamp room!"

This strategy can be extended to 23 prisoners. If the Counter follows the above rule, which is to turn the lamp off when he finds it on, and all the other prisoners follow A and B's rules, which are to turn the lamp on the first time they find it off but otherwise to do nothing, then once the Counter has seen the lamp on 22 times he knows that everyone has visited the room.

Now let's investigate what the prisoners need to do when they do not know the starting state of the lamp.

The above strategy won't work, because the first time the Counter sees the lamp on he won't be able to tell if it has been turned on by a prisoner, or if it is still in its original state.

Let's say he enters the room when the lamp is on, and that this was its original

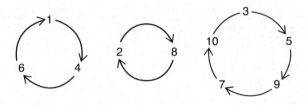

It's plain to see that there are three cycles, one of length 3, one of length 2, and one of length 5. There are more than 3.6 million permutations of 10 objects, and they can contain cycles of length 1 to length 10.

Now back to the prisoners. The strategy that they must take is the following. First, they must number themselves from 1 to 100—that is, they should order themselves into Prisoner 1, Prisoner 2, Prisoner 3, and so on.

They then must agree to abide by these rules when they enter the room:

[1] Each prisoner heads for the drawer with their number on it and opens it first. In other words, the first drawer that Prisoner 1 opens is drawer 1, the first drawer that Prisoner 2 opens is drawer 2, and so on.

[2] If a prisoner opens a drawer and it contains the name of another prisoner, say Prisoner k, the next drawer they open should be drawer k. In other words, if Prisoner 1 opens a drawer with Prisoner 32's name in it, the next drawer he should open is drawer 32. If the drawer has Prisoner 67's name in it, the next drawer he should open is drawer 67, and so on.

These two rules set each prisoner on a path that is equivalent to a permutation cycle.

Here's why. Let's imagine there are only 10 drawers and 10 prisoners. If the top row of the table on the left is relabeled "drawer number" and the bottom row is relabeled "prisoner number," the table now describes one possible arrangement of the prisoners' names in the drawers. So, for example, in drawer 1 is the name of Prisoner 4, in drawer 2 is the name of Prisoner 8, and so on.

Drawer number	1	2	3	4	5	6	7	8	9	10
Prisoner number	4	8	5	6	9	1	10	2	7	3

If Prisoner 1 abides by the rules in this strategy, he begins by opening drawer 1, in which he finds the name of Prisoner 4. So he opens drawer 4, in which he finds the name of Prisoner 6, so he opens drawer 6, in which he finds his own name. He has gone through the cycle 1 → 4 → 6 → 1, and has found his own name after opening three drawers.

We can see what happens to the other prisoners by following the cycles of this permutation (illustrated on the previous page). Prisoners 4 and 6 will also find their names after three drawers, Prisoners 2 and 8 will find their names after two, and the others will find their names after five. In other words, the number of drawers a prisoner must open to find his own name is equal to the length of the permutation cycle he finds himself in. The strategy counts him around the cycle drawer by drawer, and he finds his name when he completes the cycle.

This observation is also true when we increase the number of drawers and prisoners to 100.

If there are 100 prisoners, and each prisoner can only open 50 drawers, every prisoner will find his own name if and only if all the permutation cycles have a length of at most 50. If the length of a cycle is longer than 50, a prisoner won't be able to travel around it in only 50 drawer-openings.

The strategy works, therefore, if no permutation cycles are longer than 50. In other words, in order to work out the chances of freedom for all the prisoners—that is, of everyone finding their names—we need to calculate the chances of there being no permutation cycles longer than 50 in a random permutation of 100 objects. (In other words, we need to divide the number of permutations of 100 objects that have no permutation cycles longer than 50 by the total number of permutations of 100 objects.) The math here is too technical for a book like this one, so you'll need to trust me that the chances of there being no permutation cycles longer than 50 are just over 30 percent.

The interesting behavior of permutation cycles gives the prisoners an unexpectedly good chance of survival.

Tasty teasers

Riotous riddles

1) They are two of a set of triplets.

2) His grandchildren may have been childless, but they were still great!

3) The plane is stationary at an airport a mile above sea level.

4) She's underwater.

5) He is living in a very northerly town, in a place such as in Norway or Canada, where the sun does not rise for three months.

6) The boxer was a dog.

7) Molly is deaf.

8) She worked at the job center.

9) She witnessed my passport application.

10) He was piloting a plane heading west, flying faster than the rotational speed of Earth.

Cakes, cubes, and a cobbler's knife

GEOMETRY PROBLEMS

⑤₁ THE BOX OF CALISSONS

The solution pops off the page once you visualize the image in three dimensions. Imagine that the image is not the aerial view of a hexagonal box of diamond-shaped calissons, but instead a sideways view of a stack of small cubes inside a large cube.

It's clearer when you shade each orientation a different way, as shown below.

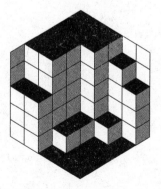

Each black calisson is the top, horizontal surface of a small cube. Each light grey calisson is a right vertical side, and each white calisson a left vertical side. If you were standing above the large cube and looked down, you would see a black 5 × 5 square made up out of all the tops of the small cubes. If you were looking at the cube from a position on the right, you would see a grey 5 × 5 square made up from all the right vertical sides of the small cubes. If you were looking at the cube from a position on the left, you would see a white 5 × 5 square made up from all the left vertical sides.

In other words, the black, grey, and white diamonds each make up a face of the large cube, and therefore each of the colors must cover the same area. Since each color contains the same number of calissons, we can deduce that the number of calissons in each orientation is the same.

Furthermore, the number of calissons in each orientation will always be the same, irrespective of the arrangement in the box.

⑸ THE NIBBLED CAKE

The cut is the straight line that goes through the center of the cake (shown below) and the center of the missing slice.

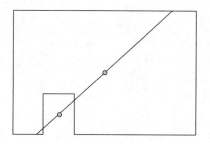

The insight needed to solve this puzzle is the realization that any straight line through the center of a rectangle divides the rectangle into two parts of equal area. Consider the cake before someone ate the rectangular slice. Any cut through the center of the cake will divide it into two equal portions. Now consider what happens with the rectangular slice taken out. If the cut goes through both the center of the cake and the center of the space left by the slice, as shown above, it will again divide the cake into two equal portions. This is because the gap left by the eaten slice is also split into two, so the area of each of the two equal portions of cake will be reduced by the same amount, and remain of equal size. Although, of course, they don't have the same shape, and one of the portions is made up of two pieces.

(53) CAKE FOR FIVE

The solution is for each of the five slices to account for the same length of the cake's perimeter. Since the perimeter of the cake is 20 units (as marked by the grid), then each of the five slices must have an edge that is four units long. So, choose a point on the perimeter and mark all the other points four units along:

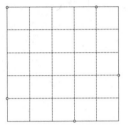

When you slice from the center of the cake to each point you are left with five equally sized slices. You may have been trying to find five slices with the same shape—but the question did not ask for that. The slices look different, but contain the same amount of cake. (And the same amount of icing, regardless of whether or not there is icing on the side of the cake.)

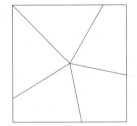

We know that the slices are of equal size because the area of each slice is either a triangle or a combination of two triangles (as shown opposite). The area of a triangle is half its base times its height. The triangles that make up the slices *all have the same height*, which is the perpendicular distance from the perimeter to the center (in this case 2.5 units). If the slice is a single triangle the base length is 4, and if the slice is two triangles, their base lengths add up to 4. So the area of all the slices is the same.

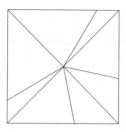

In fact, the solution works for *every possible whole number of cake slices*. If you want to slice a square cake into 7 or 9 or *n* slices, divide the perimeter of the cake into 7 or 9 or *n* equal lengths. It's essentially the same method you would use to divide a circular cake into any number of slices: Divide the perimeter into equal sections and it all works out.

54 SHARE THE DOUGHNUT

Here are two solutions with three cuts. In both of them some of the pieces are very large and some are very small. For the solution with two cuts, take away any one of the lines.

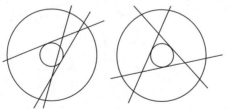

55 A STAR IS BORN

Did the title give it away? The solution creates another star out of the top triangle.

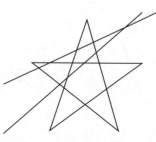

(56) SQUARING THE RECTANGLE

The rectangle is 25 cm by 16 cm, so it has an area of 400 cm². We know, therefore, that we are trying to create a 20 cm by 20 cm square. If you start the cut 20 cm along the long side you may eventually deduce the solution, which is to use a "staircase" cut. Cut the rectangle in a zigzag, as if drawing steps with 4 cm height and 5 cm width. The two pieces will now fit together perfectly.

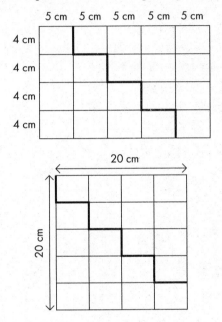

The butterfly keyboard of the 701 series IBM ThinkPad used this method. A staircase cut divided the keyboard into two pieces, and the edge between them zigzagged between the keys for 4 and 5, T and Y, H and J, and M and the comma. When the laptop was folded up, the keyboard was a square, with 4 next to J, T next to the comma, and so on. When the laptop was opened, the two pieces popped apart and fit together to make the rectangular keyboard, with 4 next to 5, T next to Y, and so on.

⑤⑦ THE SEDAN CHAIR

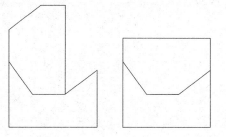

⑤⑧ FROM SPADE TO HEART

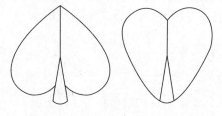

⑤⑨ THE BROKEN VASE

One straight-line cut will give you two pieces, each with a single straight side. Since the desired shape, a square, has four straight sides of equal length, you need to work out where you can make two cuts of equal length, as shown below.

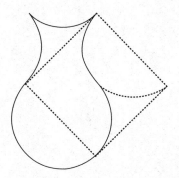

⑥ SQUARING THE SQUARE

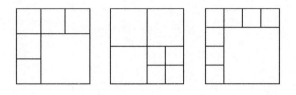

⑥ MRS. PERKINS'S QUILT

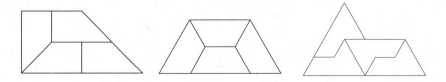

The smallest number of pieces is 11. A consolation prize if you got 12.

⑥ THE SPHINX AND OTHER REPTILES

⑥₃ ALAIN'S AMAZING ANIMALS

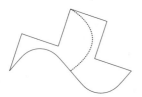

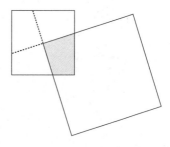

⑥₄ THE OVERLAPPING SQUARES

By extending the sides of the large square into the smaller one, we divide the small square into four. These four pieces have equal angles and side lengths, which makes them identical quarters. The area of the shaded part is therefore a quarter the area of the small square. Since the side of the small square is 2, the square's area is 4, so the shaded part's area is 1.

(65) THE CUT-UP TRIANGLE

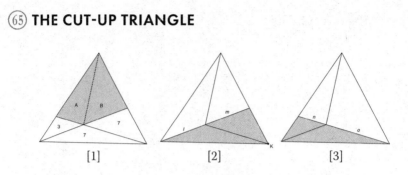

[1]　　　　　　　[2]　　　　　　　[3]

In diagram [1], I've drawn a dashed line from the intersection of the lines to the top vertex, dividing the area we want to find into A and B.

In diagram [2], one of the shaded triangles has base l and area 7, and the other has base m and area 7. These two triangles have the same height, so we can deduce that $l = m$. (The height of a triangle is the perpendicular distance from the base to the opposite vertex, K.)

But if $l = m$, the area of the other triangle with base l must be equal to the area of the other triangle with base m, since these two triangles also share the same height.

In other words, A + 3 = B.

In diagram [3], one of the triangles has base n and area 3, and the other has base o and area 7. Since these two triangles have the same height, $n = \frac{3}{7}o$.

And if $n = \frac{3}{7}o$, then the area of the other triangle with base n must be $\frac{3}{7}$ of the other triangle with base o, since these two triangles also share the same height.

In other words, $A = \frac{3}{7}(B + 7)$; or $7A = 3B + 21$.

By substituting the value for B in this equation, we get:

$7A = 3(A + 3) + 21$

$7A = 3A + 9 + 21$

$4A = 30$

$A = 7.5$

So B = 10.5

And A + B = 18

66) CATRIONA'S ARBELOS

The area is π.

The first thing to notice is that the puzzle doesn't tell us anything about the location of the vertical line within the large semicircle. We can assume that this means it doesn't matter where we place it, so let's move it to somewhere convenient, such as the center of the figure.

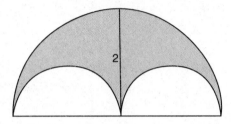

Now the problem is easy. The shaded area is equal to a semicircle with radius 2, minus two semicircles with diameter 2, or radius 1, which is a circle with radius 1. Which is $\frac{\pi 2^2}{2} - \pi 1^2 = 2\pi - \pi = \pi$.

If you're not convinced that we can move the line to the middle, then let's solve the problem another (less elegant) way. Label the diameters of the two semicircles a and b, and add in lines x and y:

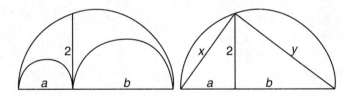

We have three right triangles here, which means we can use Pythagoras's theorem on each one. (Pythagoras's theorem states that for right triangles, the square of the hypotenuse is equal to the sum of the squares of the other two sides.) One right triangle has hypotenuse x, one has hypotenuse y, and one has hypotenuse ($a + b$), since the angle on a circle subtended by a diameter is a right angle. So:

$$x^2 = a^2 + 2^2$$
$$y^2 = b^2 + 2^2$$
$$(a + b)^2 = x^2 + y^2$$

If we combine these three equations to get rid of x and y we get:

$$(a + b)^2 - a^2 - b^2 = 8$$

The shaded area is the area of the large semicircle (radius $\frac{a+b}{2}$), minus the areas of the two smaller semicircles (which have radiuses of $\frac{a}{2}$ and $\frac{b}{2}$). This equals:

$$\tfrac{\pi}{8}(a + b)^2 - \tfrac{\pi}{8}a^2 - \tfrac{\pi}{8}b^2$$

which, with $\frac{\pi}{8}$ factored out, is $\frac{\pi}{8}\left[(a + b)^2 - a^2 - b^2\right]$

and we know the expression in [] equals 8, so the shaded area is π, regardless of the values of a and b.

⑥⑦ CATRIONA'S CROSS

The equilateral triangles cover ⅔ of the rectangle.

There are many ways to solve this problem. The method below uses no algebra. In tribute to Catriona Shearer, the puzzle's author, I'll use the "shearing" property of triangles: Two triangles with the same base and height have equal areas. (The height is the perpendicular distance from the base to the opposite vertex.) In other words, if you "shear" a triangle by moving the vertex opposite the base in a direction parallel to the base, the area does not change, because neither the base nor the height changes.

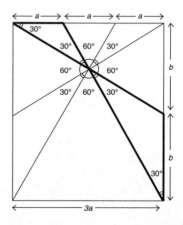

Step 1. An equilateral triangle comprises three 60-degree angles. So, the angles between the four equilateral triangles, at the point where they touch, must each be 30 degrees, since the angles must be equal, and all the angles

around the point add up to 360 degrees. Look at the bold triangle top left. The top left angle must be 30 degrees, since it adds together with the 60 degrees from the equilateral triangle below it to make a 90-degree angle. This bold triangle is therefore isosceles, which means that if the short side is a, the equilateral triangle adjacent to the short side has side length a, and the width of the rectangle is $3a$. Likewise, by considering the bold triangle bottom right, the height of the rectangle is $2b$, where b is the length of the side of the medium-sized equilateral triangle.

Step 2. If we let the equilateral triangle with side length a have area 1, then the two isosceles triangles next to it must also have area 1, since they have the same base and height. (As mentioned above, triangles with the same base and height have equal areas.) Now consider the bigger triangle that consists of these three triangles of area 1 (marked in bold in the diagram at right). It has area 3. The other triangle in bold must also have area 3, since it has the same base, b, and height.

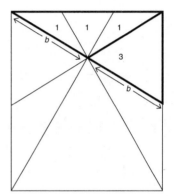

Step 3. Since we know that the medium equilateral triangle has area 3, the other triangle marked in bold below *that* must also have area 3, since it has the same base, b, and height. We can now fill in all the other areas. If you triple the base of an equilateral triangle, the area multiplies by nine. The triangles therefore cover an area of $9 + 3 + 3 + 1 = 16$ in a rectangle of area 24, which means they cover ⅔ of the total area.

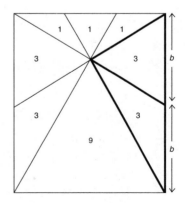

⓺⓼ CUBE ANGLE

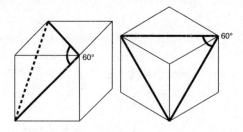

Join the two ends of the bold line to make a triangle. All three sides have the same length, so it must be an equilateral triangle, and therefore the angle is 60 degrees.

⓺⓽ THE MENGER SLICE

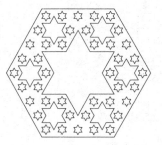

The holes turn into six-pointed stars.

⓻⓪ THE PECULIAR PEG

There are an infinite number of solutions.

Finding an object that fits through two of the holes, such as the circle and the square, is straightforward. A cylinder of height 1 and diameter 1 unit does the job. If the cylinder is sat on a table, its horizontal cross section is a circle of diameter 1. Were you to slice it vertically down a diameter, the cross section would be a square of side-length 1. This cylinder would fit in the circular hole when inserted circular end first, and in the square hole when inserted sideways.

To get the object through the triangular hole we need to carve out a triangular cross section from this cylinder that is perpendicular to both the circular and the square cross sections. One way to do this is to make two diagonal cuts, as shown below left. The object remaining will fit through the triangle when inserted in the direction of the top edge, will fit through the square when inserted perpendicular to the top edge, and through the circle when inserted base first.

This "peg," which has a perfectly triangular cross section at only one point, is the object with the largest volume that can pass through the three holes. The object with the smallest volume that can pass through the three holes is shown below right. It has the same top edge as the other shape, but every vertical cross section perpendicular to that edge is a triangle. Since the second object can be carved out of the first, it is possible to create an infinite number of different-sized pegs that are trimmed from the first object but not yet reduced to the second one.

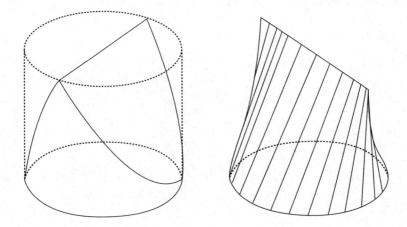

The second one looks a bit like an Afghan *karakul* hat.

(71) THE TWO PYRAMIDS

The solid has five sides, because when you put the pyramids together you get a shape like a slice of cheese. Two of the faces on the triangular-based pyramid are co-planar with two of the faces on the square-based pyramid.

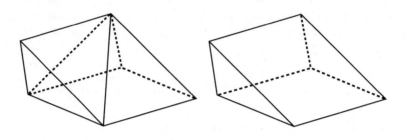

The diagram of the two square-based pyramids helps us visualize the solution. A line drawn between the tops of the two pyramids, as below, creates the outline of a triangular-based pyramid (a tetrahedron) slotted perfectly between them. One face of this tetrahedron coincides perfectly with the pyramid on the left, and another face coincides perfectly with the pyramid on the right. The sides of the two pyramids and the tetrahedron that face the reader (and that face away from the reader) all lie on the same plane.

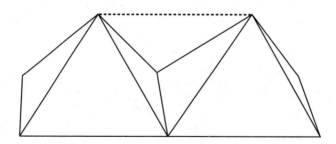

(72) THE ROD AND THE STRING

Imagine that the rod is a cylinder—a cardboard paper-towel-roll cylinder, say. Cut a straight line from one end of the roll to the other, between the points where the ends of the string are. When you unroll the cylinder and place it on a flat surface, you will get a rectangle that is 12 cm by 4 cm, as illustrated here, which can be divided into four equally sized, 3 cm × 4 cm rectangles according to the points where the string crosses the cut edge.

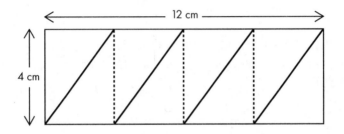

The diagonal of one of these 3 cm × 4 cm rectangles is the hypotenuse of a right triangle. Here we use the only piece of technical math needed, Pythagoras's theorem, which states that the square of the hypotenuse of a right triangle is equal to the sum of the squares of the other two sides. Using Pythagoras, the hypotenuse has length $\sqrt{(3^2 + 4^2)} = \sqrt{(9 + 16)} = \sqrt{25} = 5$. The total length of the string is equal to four hypotenuses, which is $4 \times 5 = 20$ cm.

(73) WHAT COLOR IS THE BEARD?

The beard is white. Everyone knows that Santa Claus lives at the North Pole! (For those who do not believe in Santa, or claim he resides at an alternative address, any bearded person traversing the extreme north will also likely have a beard colored white by snow and frost.)

There are an infinite number of places in the world from which a path 10 miles due north, then 10 miles due west, then 10 miles due south takes you back to your starting point. One is the South Pole, where this path maps out a triangle. The question, however, ruled this out as an answer.

The other places are all very close to the North Pole.

Consider the circle of latitude with a circumference of exactly 10 miles (which is only a few miles south of the North Pole). Let your starting point be any point 10 miles south of this circle, as illustrated below. You start your trek, and after 10 miles due north, you reach the circle at, say, point A. The 10 miles due west that you now travel is, in fact, one circumnavigation of the circle, which returns you to A. Traveling due south now for the final leg returns you to your starting position. Equally, consider the circle of latitude that is exactly 5 miles long, or 2 miles long, or indeed any distance that divides 10 miles into a whole number. If you were to start 10 miles south of any one of these circles, all of which are very close to the North Pole, you would also return back to your starting point, because a 10-mile walk due west on any of these circles will return you to your starting point after some number of laps around.

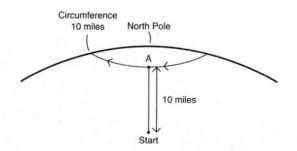

⑦④ AROUND THE WORLD IN 18 DAYS

Phileas Fogg arrives back on October 19th.

If you travel around the world *eastward*, the sun sets and rises sooner, so you experience shorter days. After 18 sunsets and sunrises (which you experience as 18 days), only 17 days will have passed at your point of origin.

Indeed, Jules Verne used this quirk of circumnavigation in the plot of his novel *Around the World in Eighty Days*. Phileas Fogg bets that he can travel around the world in 80 days. He counts 80 days on his journey, but due to a missed train arrives back in London five minutes after the deadline. He thinks he has lost the wager. In fact, he is a day early, and he wins it.

⑦⑤ A WHISKEY PROBLEM

This is a beautiful puzzle because it can be solved with a simple insight. When I suggested that you might want to drink some more, but not too much, I was giving you a clue.

When the upright bottle contains whiskey to a height of 14 cm, there is 13 cm above it with no whiskey.

If you were to drink 3 cm of the whiskey, leaving 11 cm, there would now be 16 cm above it with no whiskey.

The upturned bottle would now contain 19 − 3 = 16 cm of whiskey. In other words, the volume of whiskey in the bottle is equal to the volume of the bottle that is empty, meaning that the bottle is half-full.

So, the whiskey in the bottle is equal to half the bottle's volume plus 3 cm of whiskey.

The full bottle is 750 cm^3, so half the bottle is 375 cm^3. The only calculation we need to do is to find the volume of 3 cm of whiskey, which is $\pi r^2 \times 3$ cm, where r is the radius of the bottle. Since the diameter is 7 cm, the radius is 3.5 cm, so the volume of 3 cm is:

$3.14 \times (3.5)^2 \times 3 = 115$ cubic cm (approximately).

The whiskey left in the bottle = 375 + 115 = 490 cubic cm (approximately).

Cheers!

Tasty teasers

Pencils and utensils

1) Step 1: Rotate your right hand so that your four right fingers are between the thumb and index finger of your left hand.

Step 2: Place your right index finger on the pencil and twist it around in a clockwise direction, while bringing your left thumb around your right thumb.

2) A magnetic rod will have positive charge at one end, negative charge at the other, and no charge in the middle. So, move one of the rods so that its end is touching the middle of the other rod, as if to make a T. If there is no attraction between the rods, the "stem" is not the magnet, but if there is an attraction, the stem is the magnet.

3) The dentist puts on both pairs of gloves to treat the first patient, and then takes the outer pair off. She treats the second patient with the gloves that are still on. The assistant turns the gloves that the dentist is no longer wearing inside out, and puts them back on the dentist, who is still wearing the first pair of gloves. The two contaminated sides are in contact, leaving the sterile side on the outside, so the dentist can treat the third patient.

4)

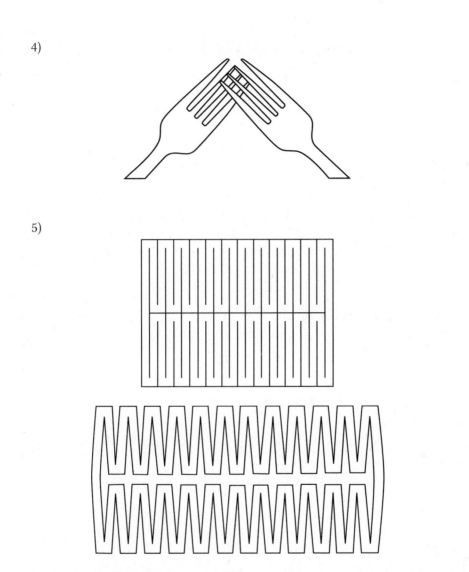

5)

If you cut the postcard in this zigzag pattern, it will open up to become a loop large enough for a person to step through.

A wry plod

WORD PROBLEMS

⑦⑥ THE SACRED VOWELS

[1] Ellen the jezebel chews sweet preserves then belches beer.
[2] Orthodox monk Otto wolfs down two bowls of pork wonton.
[3] An Arab lass gnaws at a lamb shank and has jam tarts as a snack.
[4] Bud's drunk chum glugs rum punch, upchucks hummus brunch, slumps.
[5] This rich dish is figs in icing, which I finish whilst swigging gin.

⑦⑦ WINTER REIGNS

The first letter of each new word is the same as the last letter of the preceding word.

⑦⑧ FIVE DEFT SENTENCES

[1] All of these words contain three consecutive letters of the alphabet, and the final one has four: Deft Afghans hijack somnolent understudies. Only a few words in English have this property.
[2] The sentence is a palindrome, meaning that the letters read the same forward as they do backward.
[3] The first sentence is formed exclusively from letters on the top row of a keyboard. The second uses only letters from the middle row, and the third has only letters from the bottom row.
[4] The first word has one letter, the second word has two, the third three, and so on, with the eleventh word having eleven letters. This constraint is what's known as a "snowball sentence." This example, from *Language on Vacation*, by Dmitri

Borgmann, has 20 words, although I didn't use the second part since it rather gives the game away:

I do not know where family doctors acquired illegibly perplexing handwriting; nevertheless, extraordinary pharmaceutical intellectuality, counterbalancing indecipherability, transcendentalizes intercommunications' incomprehensibleness.

[5] This obeys Oulipo's "prisoner's constraint," which forces you to use only letters without ascenders or descenders. In other words, only the letters a, c, e, i, m, n, o, r, s, u, v, w, x, and z. The other letters all extend either above or below the height of these letters. It's called the prisoner's constraint because if you were in prison with only a single piece of paper, you would be able to fit more lines of text on the page.

⑦⑨ THE CONSONANT GARDENER

S	T	R	E	N	G	T	H	S				
A	B	R	A	C	A	D	A	B	R	A		
V	E	R	I	S	I	M	I	L	I	T	U	D	E	S
F	A	C	E	T	I	O	U	S	L	Y		
A	M	B	I	D	E	X	T	R	O	U	S	L	Y

⑧⓪ KANGAROO WORDS

hen	fiction	part	sated
taint	noble	me	
dead	tutor	rain	

(81) THE TEN-LETTER WORDS

Each word contains four different letters, of which one appears once, one twice, one three times, and one four times. Words in which one letter occurs once, a second letter occurs twice, a third letter occurs three times, and so on, are called "pyramid words."

(82) TEN NOTABLE NUMBERS

[1] Four
[2] Eighty-four
[3] Eighty-eight
[4] Forty
[5] One
[6] One thousand and eighty-four
[7] Twenty
[8] One hundred and one
[9] One billion
[10] One octillion

(83) THE QUESTIONS THAT COUNT THEMSELVES

[1] 73
[2] 79
[3] 75
[4] 93
[5] 168

(84) THE SEQUENCE THAT DESCRIBES ITSELF

SIX, ONE, EIGHT, FIVE, TWO, SEVEN, FOUR, TEN, ONE, TWO . . .

Here's how we get there. We started with:

SIX, ONE / E / IGHT FIVE / . . .
   6   1   8

We know that the next E must appear at the end of a chunk of five:

SIX, ONE / E / IGHT FIVE / * * * * E/
   6   1   8   5

We can eliminate numbers with an E before the fifth term, such as ONE, THREE, FIVE, SEVEN, EIGHT, NINE, TEN, and so on. This leaves us with TWO, FOUR, and SIX. Let's try TWO, the lowest number, since the question tells us we must choose the lowest number that fits. The TWO gives us the length of the next chunk:

SIX, ONE / E / IGHT FIVE / TWO * E/ *E/
   6   1   8   5   2

The only number that fits here is SEVEN:

SIX, ONE / E / IGHT FIVE / TWO SE/ VE/ N ***** E /
   6   1   8   5   2   7

Again let's choose the lowest number that fits, which is TWO:

SIX, ONE / E / IGHT FIVE / TWO SE/ VE/ N TWO ** E / * E
   6   1   8   5   2   7   2

However, no number fits the gap, so we can eliminate TWO as wrong. Let's try FOUR:

SIX, ONE / E / IGHT FIVE / TWO SE/ VE/ N FOUR * E / * ** E
   6   1   8   5   2   7   4

The lowest number that works is TEN:

SIX, ONE / E / IGHT FIVE / TWO SE/ VE/ N FOUR TE / N ** E/ *********E

    6   1   8   5   2   7   4   10

We complete the sequence by continuing in this vein.

85 SEXY LEXY

One billion: It has ten letters, and 1,000,000,000 has ten digits.

86 LETTERS IN A BOX

A	B	S	C	O	N	D	E	D		

B	I	F	U	R	C	A	T	E	D	

H	E	A	D	A	C	H	E			

M	O	N	I	K	E	R				

D	I	S	L	I	K	E	A	B	L	E

G	L	O	W	W	O	R	M			

H	I	T	C	H	H	I	K	E	R	

M	I	C	R	O	O	R	G	A	N	I	S	M

87 WONDERFUL WORDS

[1] Wholesome.
[2] It is the longest English word that reverses to make a different word: in this case, desserts.
[3] Natal, Nice, and Reading are all places.

[4] "Yes" becomes "Ayes."

[5] Short.

[6] "Nowhere" becomes "now here."

[7] "Queue," which has the same pronunciation as "Q." Or "aitch," which has the same pronunciation as "H."

[8] Because *end* does indeed begin with an "e."

[9] The adjectives are synonymous with mercurial, venereal, earthy, martial, jovial, and saturnine, which each come from the names of the planets, in order of their distance from the Sun: Mercury, Venus, Earth, Mars, Jupiter, and Saturn.

[10] A mailbox.

(88) LIFE SENTENCES

[1] Buffalo buffalo Buffalo buffalo buffalo buffalo Buffalo buffalo.

The word "buffalo" has three meanings: as a city, as an animal, and as a verb. One way to make the sentence less confusing is to clearly distinguish between these meanings by rewriting the city name as "from the city of Buffalo," and to substitute the animal for "bison" and the verb for "intimidate." So we get:

Buffalo buffalo (i.e., bison from the city of Buffalo) Buffalo buffalo buffalo (i.e., the ones that are intimidated by bison from the city of Buffalo) buffalo Buffalo buffalo (i.e., intimidate bison from the city of Buffalo).

In other words, the bison from Buffalo that are intimidated by other bison from Buffalo, themselves intimidate bison from Buffalo.

[2] John, where James had had "had," had had "had had"; "had had" had had the teacher's approval.

Here's a context in which this makes sense. A teacher asked John and James to write an English sentence in the pluperfect tense. John wrote the sentence "The woman had had a baby." James wrote "The woman had a baby."

[3] Sunday.

The day after tomorrow is in two days' time. So when the day after tomorrow is yesterday, this date—"today"—will be in three days' time.

Likewise, the day before yesterday is two days ago. So when the day before yesterday is tomorrow, this date—"today"—is three days ago.

If these two dates—three days ago, and in three days' time—are equidistant from Sunday, then today must be Sunday.

89 IN THE BEGINNING (AND THE MIDDLE AND THE END) WAS THE WORD

The word is "only."

90 LOOKING AT LETTERS

[1] It's an E sideways.

[2] It's the alphabet with a twist! Or with successive quarter-turns. Listing the letters backward from Z, the Z has been rotated 90 degrees (either clockwise or counterclockwise). The Y has not been rotated, the X has been rotated by 90 degrees (either clockwise or counterclockwise). The fourth letter is a W that has been rotated 180 degrees. The next symbol would be a V turned by 90 degrees, which is either < or >.

[3] Turn the page sideways and you will read ONION, NOON, ZOO, and NUN.

91 A MATTER OF REFLECTION

The answer is HID, which is the only word in the text made from letters that look the same when reflected horizontally.

⑨² THE BLANK COLUMN

The phenomenon occurs for all sentences, provided the text is aligned with the left margin and words are not broken on the right.

Let's consider the phrase in the question: "Perilous problems for puzzle lovers!" In the second line, which starts with "problems for," these words are immediately followed by "puzzle lovers!" In other words, the first five words of the line are precisely the five words in "Perilous problems for puzzle lovers!"—just not in the correct order.

Likewise, in the third line, which starts "for puzzle lovers," the first five words are *also* the five words in "Perilous problems for puzzle lovers!"—just not in the correct order. In fact, whichever word starts the line, the first five words will *always* be the five words in "Perilous problems for puzzle lovers!" If the sentence starts with the same five words, it follows that the fifth word will always end at the same position along, and therefore immediately after the fifth word there will be a blank space.

We can generalize this to any sentence S with *n* words. Whatever word begins the line, the first *n* words of that line are precisely the words in S. It follows that the *n*th word must always end in the same position, so immediately after the *n*th word (or the punctuation point after that word) is a space.

⑨³ WELCOME TO THE FOLD

94 MY FIRST AMBIGRAM

Scott Kim gives these as possible solutions:

95 BOXED PROVERBS

[1] Look before you leap.

[2] A stitch in time saves nine.

[3] A rolling stone gathers no moss.

[4] Silence is golden.

[5] Beggars can't be choosers.

[6] All's well that ends well.

[7] Beauty is only skin deep.

[8] All roads lead to Rome.

[9] Better safe than sorry.

[10] You can't judge a book by its cover.

96 NMRCL ABBRVTNS

[1] Winter, spring, summer, and fall are the 4 seasons.

[2] Africa, Asia, Australia, Antarctica, Europe, North America, and South America are the 7 continents.

[3] Sneezy, Grumpy, Dopey, Bashful, Sleepy, Doc, and Happy are the 7 dwarfs.

[4] The alphabet has 26 letters.

[5] A guitar has 6 strings.

[6] There are 52 cards in a deck.

[7] A spider has 8 legs.

[8] A chessboard has 64 squares.

[9] A golf course has 18 holes.

[10] A 1 followed by 6 zeros is a million.

(97) THE NAME OF THE FATHER

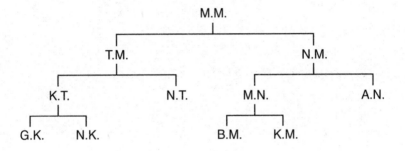

(98) TELLING THE TIME IN TALLINN

From the clock faces we can deduce right away that *üks* is 1, *kaks* is 2, and *viis* is probably 5. (Since it's likely that *minutit* means "minute" and *läbi* means "past.") The words *veerand* and *kolmveerand* seem to refer to a quarter hour and three quarters of an hour, so *kolm* is probably 3.

What's confusing now is that the expression for 1:15 has *kaks*, the word for 2, in it, and the expression for 10:45 has *üksteist* in it, which looks like it must be 11. (Because the only other similar expression in the question is *kaksteist*, which would be 12.)

The deductive leap is to realize that, in Estonian, a "quarter past the hour" is instead described as "a quarter of an hour on the way to the next hour," "half past the hour" is described as "half an hour on the way to the next hour," and "three quarters past the hour" is described as "three quarters of an hour on the way to the next hour." In seeing that both *pool neli* and *veerand neli* are expressions, we can deduce that *neli* is 4, which means that the only number left, *üheksa*, must be 9. So the answers are:

Estonian to numbers:

    [1] 9:25   [3] 2:30   [5] 6:35

    [2] 11:45   [4] 3:15

Numbers to Estonian:

[1] Veerand viis.   [3] Pool kaksteist.   [5] Pool üks.

[2] Kolmveerand üheksa.   [4] Viis minutit seitse läbi.

⑨⑨ COUNTING IN THE RAINFOREST

1 = aroke

2 = měña

3 = měña go aroke

4 = měña go měña

5 = ãěmãěmpoke

6 = ãěmãěmpoke go aroke

7 = ãěmãěmpoke go měña

8 = měña měña měña měña

9 = ãěmãěmpoke měña go měña

10 = tipãěmpoke

From [2], there are two numbers such that the sum of their squares is at most 10. Each of these two numbers must be either 1, 2, or 3, so we can infer that *aroke* and *měña* are 1, 2, or 3, and that *ãěmãěmpoke* is 5 or 10.

From [4] we can see that *měña* cannot be 1, so *ãěmãěmpoke* must be 5, since if it were 10 its product with *měña* would be greater than 10. If *ãěmãěmpoke* is

5, *aroke* and *měňa* are either 1 or 2. But since *měňa* is not 1, it must be 2, which means that *aroke* is 1; thanks to [4] *tipăěmpoke* is 10.

We still have to account for 3, 4, 6, 7, 8, and 9. From [3], the number which is squared, *ăěmăěmpoke go aroke*, cannot be 3, since the only two numbers that multiply to equal 9 are 1 and 9, and 1 is already taken. Likewise, we can eliminate 4, 7, 8, and 9, so *ăěmăěmpoke go aroke* must be 6, and the two numbers on the other side of the equation are 4 and 9 (since $4 \times 9 = 36$). The expression on the left is *měňa go měňa*, and the one on the right is *ăěmăěmpoke měňa go měňa*—in other words, the same expression but including the word for 5. So it makes sense for *měňa go měňa* to be 4 and for *ăěmăěmpoke měňa go měňa* to be 9.

Now to [1]: *měňa měňa měňa měňa* = 12 − 4 = 8. The missing numbers now are 3 and 7. We build 3 (which is 2 + 1) by analogy to 6 (which is 5 + 1) and we build 7 (which is 5 + 2) by analogy to 4 (which is 2 + 2).

The solution reveals an underlying structure that is a mixture of base two and base five.

(100) CHEMISTRY LESSON

This question requires intelligent guesswork as well as pure logic.

The names have four possible prefixes: *hept-, prop-, but-,* and *eth-;* and three possible suffixes: *-ane, -ene,* and *-yne.*

Likewise, there are four types of carbon: C_2, C_3, C_4, and C_7. A sensible guess is that the carbon numbers correspond to the prefixes, and that C_7 is *hept-,* since *hept-* is a prefix that in other contexts, such as "heptagon" and "heptathlon," means 7. Two formulas contain C_3 and two C_4, and likewise two names contain *prop-* and two contain *but-,* but only one formula contains C_2 and only one name has *eth-,* so we can deduce that C_2 must be *eth-.*

So far we have:

Heptene: C_7H_{14}

Ethyne: C_2H_2

There are five types of hydrogen: H_2, H_4, H_6, H_8, and H_{14}, but only three suffixes, so suffixes do not correspond to the type of hydrogen.

state. The Counter will turn it off, as he must, but now for him to be sure that everyone has been in the room he must wait to see the lit lamp another 22 times. In other words, he must wait until he has turned the lamp off 23 times. However, the rule "wait until I see an on-lamp 23 times" cannot be a solution to the problem, because if the lamp started in the "off" state, it's impossible for him to see the lamp lit 23 times.

The way around this problem is to keep the same rules for the Counter but to tweak the rules for the other prisoners by insisting that they turn the lamp on twice. The rules become:

If this is the *first or second time* you have seen the lamp off, turn it on.

In all other cases, do nothing.

The Counter can now be sure that once he has seen the lit lamp 44 times, everyone has been in the room.

If the lamp was off at the start, every other prisoner would have gone in twice. If the lamp was on at the start, every other prisoner would have gone in twice except for one, who would have gone in once. It's possible that all the prisoners had been in the room before the Counter saw the 44th on switch, but he can only know for certain when he sees the "on" switch that 44th time.

(50) THE ONE HUNDRED DRAWERS

Before we get to the solution, here's an introduction to the math of permutations. It's going to make the prisoners' strategy much easier to understand. Let's say we have 10 objects and we want to reorder them. One way to describe this reordering is in a table:

Initial position	1	2	3	4	5	6	7	8	9	10
New position	4	8	5	6	9	1	10	2	7	3

An easy way to understand the pattern described in the table is to represent it visually. In the table 1 → 4, 4 → 6 and 6 → 1. This forms a loop, or "permutation cycle." The table can be illustrated thus:

What else could the suffixes refer to? Let's consider the ratios between the numbers. With heptene, the H number is double the C number. There is only one other -*ene*, and only one other formula in which the H number is double the C number. So let's go with the idea that -*ene* means H, which is double C. So:

Butene: C_4H_8

Which leaves us with one remaining C_4, which must be the other *but*-.

Butyne: C_4H_6

Now we seem to have run into a problem. If the suffix is based on the ratio between C and H, why do ethyne (C_2H_2) and butyne (C_4H_6) have the same suffix but different ratios? The insight here is to realize that H-numbers are always even, so the ratio might have something to do with being *half* the H-number. This would give ethyne a C:($\frac{H}{2}$) ratio of 2:1, and butyne a C:($\frac{H}{2}$) of 4:3. In both cases C = ($\frac{H}{2}$) + 1, which suggests that -*yne* occurs when C = ($\frac{H}{2}$) + 1.

The two remaining formulas have a C:($\frac{H}{2}$) ratio of 3:2 and 3:4, which would make the former propyne and the latter, by elimination, propane. So:

Propyne: C_3H_4

Propane: C_3H_8

Tasty teasers
Bongard bafflers

1) Left: sets of three colinear points. Right: no sets of three colinear points.

2) Left: circles on different curves. Right: circles on the same curves.

3) Left: more circles inside than outside. Right: fewer circles inside than outside.

4) Left: black region widens in the direction of the opposite side of the shape. Right: black region narrows in the direction of the opposite side of the shape.

Sleepless nights and sibling rivalries

PROBABILITY PROBLEMS

(101) BETTER THAN HALF A CHANCE

Place a single dark chocolate cookie in one jar, and the other 99 cookies in the other jar. With this arrangement, if you open one of the jars you will have a 100 percent chance of choosing a dark chocolate cookie, and if you open the other you will have a $\frac{49}{99} = 49.49$ percent chance. So on average you have almost a 75 percent chance of getting your preferred flavor.

(102) SINGLE WHITE PEBBLE

If the urn contains an even number of pebbles the chances of you or your friend picking the white pebble are equal. Going first conveys no advantage. You might as well separate the pebbles into two equally sized groups. The white pebble has a 50 percent chance of being in either group.

In fact, the probability of picking the white pebble at any stage is the same. If there are n pebbles in total, you have a $\frac{1}{n}$ chance of getting the pebble on your first pick. The chances of getting it on your second pick are the chances of *not* getting it on the first pick, which are $\frac{(n-1)}{n}$, multiplied by the chances of getting it on the second pick, which are $\frac{1}{(n-1)}$. This gives you a $\frac{1}{n}$ chance.

If the urn contains an odd number of pebbles, it's worth going first because you get one more attempt.

⑩³ THE JOY OF SOCKS

Four socks in total: two red and two blue. The minimum number of socks you need to take out to ensure you have either two of the same or two of different colors is three.

⑩⁴ LOOSE CHANGE

If you take 20 coins out of your pocket and at least one is a nickel, this implies that there must be at least seven nickels among the 26 coins. (If there were six or fewer nickels you would not be guaranteed to get one among the 20 coins you take out.) Likewise, if you take out 20 coins and at least two are dimes, you must have at least eight dimes in your pocket, and if you take 20 coins out and at least five are quarters, you must have at least eleven quarters.

If you have at least seven nickels, at least eight dimes, and at least eleven quarters, you must have exactly those amounts, since 7 + 8 + 11 = 26. So the total money in your pocket is 11 × 25¢ + 8 × 10¢ + 7 × 5¢ = $2.75 + $0.80 + $0.35 = $3.90.

⑩⁵ THE SACK OF POTATOES

There are 11 potatoes in the sack, which means that there are 2^{11} = 2,048 ways that we can take a number of potatoes from the sack. We could take no potatoes. Or we could take one potato, two potatoes, three potatoes . . . stop me now before I break into song.

The 2,048 combinations of potatoes that we can take from the sack will have weights that vary between 0 g and 2,000 g. Since there are more possible combinations than there are weights (2,048 > 2,000), there must be at least two combinations of potatoes that weigh the same number of grams (rounding to the nearest gram). Let's call these two combinations A and B.

If A and B do not have any potatoes in common, we can remove each of these combinations from the sack, and the weight of the pile made from A will differ from the weight of the pile made from B by less than 1 g.

If A and B do have potatoes in common, we can simply remove these common potatoes, and what's remaining in pile A and what's remaining in pile B will still have weights that differ by less than 1 g.

(106) THE BAGS OF CANDIES

Fifteen candies. Place a candy in the first bag, then place this bag, with another candy, in the second bag, then place these two bags, with another candy, into the third bag, and so on. The fifteenth bag will contain all 15 candies and all the other bags too. Stashed in this way, each bag will contain a different number of candies (and bags).

(107) A STRATEGY FOR THE DISPLACEMENT OF IMPROPER THOUGHTS

The chance that the ball in the bag will be white is $\frac{2}{3}$, or 66.7 percent.

We solve this problem by considering equally likely outcomes.

The ball in the bag is either black or white. When a white ball is placed in the bag, there are two equally likely possibilities: Either the bag now has a black and a white ball, or it now has two white balls. When a ball is taken out of the bag at random, there are now four equally likely possibilities, as described in this table. (When there are two white balls, one is called white$_1$, and the other is white$_2$.)

Balls in the bag	Ball taken out	Ball still in bag
black, white	black	white
black, white	white	black
white$_1$, white$_2$	white$_1$	white$_2$
white$_1$, white$_2$	white$_2$	white$_1$

There are three possibilities in which a white ball has been removed, and in two of them the ball inside the bag is white. So the chance of a white ball being in the bag is 2 in 3, or 66.7 percent.

(108) BERTRAND'S BOX PARADOX

The chance that the other counter in the box is black is $\frac{2}{3}$, or 66.7 percent.

Like the previous problem, we need to consider all equally likely outcomes.

When you open a box at random and choose a counter at random, every counter has an equal chance of being chosen. If the counter is black, you have chosen one of the three black counters. The crucial insight here is that each black counter is an equally likely choice. I've labeled the black counters A, B, and C.

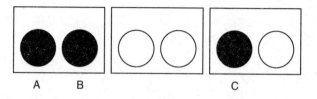

If you removed A, the other counter in the same box is black.

If you removed B, the other counter in the same box is black.

If you removed C, the other counter in the same box is white.

In other words, in two out of three equally likely outcomes, the remaining counter is black.

So the chance of the remaining counter being black is 2 in 3, or 66.7 percent.

⑩ THE DICE MAN DIET

Monday.

You have a 1 in 6 chance of eating your first dessert today. In order for your first dessert to be eaten tomorrow, you must not eat dessert today (a 5 in 6 chance) *and* roll a 6 tomorrow (another 1 in 6 chance). A ⅚ chance and a ⅙ chance combined is less than a ⅙ chance, so it's more likely that your first dessert will be today than tomorrow. Following that logic, "first dessert" on the day after tomorrow becomes even less likely than "first dessert" tomorrow, and so on.

⑩ DIE! DIE! DIE!

The bet is not in your favor. In fact, this is the casino game chuck-a-luck, often played with the three dice in a spinning cage.

To show that the house always wins, imagine that six players bet $100 on each of the six possible outcomes. The house takes in $600.

If all three of the dice land on the same face, the house pays back $400 ($300 to the winner plus the stake).

If two dice land on the same face, the house pays back $500 ($200 to one winner, $100 to another winner, plus their two stakes).

If all three land on different faces, the house pays back $600 (three $100 prizes plus the three original stakes).

In other words, the house never loses. Which means that, overall, over time, the gamblers never win.

⑪ THE PHONY FLIPS

The second sequence is the one I made up.

Let's look at the similarities: Both sequences contain an equal number of heads and tails, which is what we might expect from a random set of flips.

Now to the differences. Look at the largest runs in each sequence. The first sequence has a run of five *T*'s and a run of four *H*'s. The largest runs in the second sequence, however, are three *T*'s and three *H*'s. While a run of five *T*'s might seem unlikely in a sequence of 30 flips—and therefore evidence of human interference—in fact, an occurrence is more likely than not. Randomness throws up this type of cluster, or "coincidence," all the time.

The second sequence gives itself away as man-made if you count the number of times the flips alternate between heads and tails. If all the flips are random, you would expect a head (or tail) to be followed by a tail about 50 percent of the time and followed by a head about 50 percent of the time. So, in a run of 30 flips, you would expect the results to alternate about 14 or 15 times. In the first sequence, the results alternate between heads and tails 14 times, while in the second sequence the results alternate 18 times. The second sequence therefore appears more biased, and thus less random, than the first.

⑫ JUST FOUR KIDS

At first glance, you might think that half boys and half girls is more likely. Each child has a 50/50 chance of being a boy or a girl, so with four siblings you would expect, on average, two to be boys and two to be girls. Sloppy thinking!

Let's look at the 16 equally probable combinations of four siblings (where B is a boy and G is a girl):

BBBG	GGGB	**GGBB**	**GBBG**
BBGB	GGBG	**BBGG**	**BGGB**
BGBB	GBGG	**GBGB**	BBBB
GBBB	BGGG	**BGBG**	GGGG

Once you add them up, you'll see that only **six** possible combinations have two boys and two girls, but eight have three of one and one of the other.

So the most likely outcome will be three of one and one of the other.

⑪⑬ THE BIG FAMILY

Mrs. Brown's strategy is likely to result in a smaller family.

Let G = a girl and B = a boy. There are four equally likely "doublets" of two children born one after the other: BB, BG, GG, and GB.

The question states that Mr. Brown wants to stop when they get a BB, and that Mrs. Brown wants to stop when they get a GB.

In other words, the problem boils down to working out whether BB or GB is more likely to come first in a random sequence of Bs and Gs.

In a family of two children, the sequence BB and BG appear with equal likelihood.

However, in a family of more than two children, GB is *more likely* to appear before BB. This is because in any sequence of Bs and Gs, a BB is always preceded by a GB except when the first two children are BB. In other words, as soon as a girl is born, you are guaranteed that GB will come before BB.

Since there is only a 1 in 4 chance that the first two children are BB, the chance that GB appears first is 3 in 4.

In other words, when it gets to the point that one of them wants to stop having children, it is more likely to be Mrs. Brown (75 percent chance) than Mr. Brown (25 percent chance).

⑪⑭ PROBLEMS WITH SIBLINGS

[1] ⅓

Problems involving probabilities are often easier to understand when expressed as expected frequencies. Imagine that 4,000 two-children families are chosen at random. We'd expect the following frequencies for each equally likely combination:

Older child	Younger child	Frequency
Boy	Girl	1,000
Girl	Boy	1,000
Boy	Boy	1,000
Girl	Girl	1,000

In all occurrences of boy-girl, Albert will circle the first line. That's 1,000 cases.

In half of the occurrences of boy-boy, Albert will circle the first line. That's 500 cases.

So, the top line is circled in 1,500 cases, and in only 500 of them are both children boys. The chance of two boys is therefore 1 in 3.

[2] No.

Here we have a lovely paradox, which captures one of the reasons that this type of problem causes lots of anguish and disagreement. We know from above that Albert has a 1 in 3 chance of having two children. All the journalist has done is communicate what Albert wrote on the form. So if readers were aware of the rules that governed how Albert filled in the form, they would deduce that Albert has a 1 in 3 chance of having two boys. Yet if readers assume that Albert has been randomly selected from all two-children families in which the older child is a son, then we know from the preamble to the puzzle that the chance of his having two boys is 1 in 2.

The moral is: In order to avoid any ambiguity we need to know how the information presented was arrived at.

[3] ½

If Beth is thinking of the older child, there is a 1 in 2 chance the other child is a girl. If she is thinking of the younger one, there is also a 1 in 2 chance that both are girls. So the overall chance must be 1 in 2.

[4] No it doesn't, because the three possible combinations—GG, GB, and BG—are all still equally likely.

[5] ½

Remarkably, this situation produces a different answer from the previous one. Imagine I asked this question 100 times. On 50 occasions I would have asked "Is your older child a girl?," and a yes to this answer would have given 25 instances in which two children are girls. On the remaining 50 occasions I would have asked "Is your younger child a girl?" and a yes would have given another 25 instances of two girls. So, in 50 of 100 instances there are two girls, meaning that the chances are 1 in 2.

115 THE GIRL BORN IN AN EVEN YEAR

Let's approach this problem by looking at frequencies, as we did in the answer to the previous question. Imagine selecting 400 two-children families at random. We'd expect the following frequencies:

Older child	Younger child	Frequency
Boy	Girl	100
Girl	Boy	100
Boy	Boy	100
Girl	Girl	100

We know that babies are equally likely to be born in odd or even years. If $girl_{ODD}$ and $girl_{EVEN}$ refer to girls born in odd and even years, the table becomes:

Older child	Younger child	Frequency
Boy	$Girl_{ODD}$	50
Boy	$Girl_{EVEN}$	**50**
$Girl_{ODD}$	Boy	50
$Girl_{EVEN}$	Boy	**50**
Boy	Boy	100
$Girl_{EVEN}$	$Girl_{EVEN}$	**25**
$Girl_{EVEN}$	$Girl_{ODD}$	**25**
$Girl_{ODD}$	$Girl_{EVEN}$	**25**
$Girl_{ODD}$	$Girl_{ODD}$	25

The total number of families with at least one $girl_{EVEN}$ is **50 + 50 + 25 + 25 + 25 = 175.** Of these families, 75 of them have two girls.

So the probability of having two girls, given that one is a $girl_{EVEN}$, is $^{75}/_{175} = \frac{3}{7}$, or about 43 percent.

Now to the Tuesday–boy problem, in which a parent has a boy born on a Tuesday. We can draw up a similar table of frequencies. Imagine selecting 196 two-children families at random. (It will become clear why I chose that particular number.) We'd expect the following frequencies:

Older child	Younger child	Frequency
Boy	Girl	49
Girl	Boy	49
Boy	Boy	49
Girl	Girl	49

Each equally likely combination has a frequency of 49. When dealing with days of the week it is helpful to have a number divisible by 7. Now if boy_T is a boy born on a Tuesday and boy_{NT} is a boy born on any other day, we can expand the table to:

Older child	Younger child	Frequency
Boy$_T$	Girl	**7**
Boy$_{NT}$	Girl	42
Girl	Boy$_T$	**7**
Girl	Boy$_{NT}$	42
Boy$_T$	Boy$_T$	**1**
Boy$_{NT}$	Boy$_T$	**6**
Boy$_T$	Boy$_{NT}$	**6**
Boy$_{NT}$	Boy$_{NT}$	36
Girl	Girl	49

The number of families with a boy born on a Tuesday is **7 + 7 + 1 + 6 + 6 = 27.** Of these families, 13 have two boys, so the probability of having two boys, given that one is born on a Tuesday, is $^{13}\!/_{27}$, which is about 48 percent.

Let's summarize what we have discovered.

When a two-child family has "at least one boy," the chances of two boys is 33 percent.

When a two-child family has "at least one boy who was born in an even year," the chance of two boys is 43 percent.

When a two-child family has "at least one boy who was born on a Tuesday," the chance of two boys rises to 48 percent.

As more details are specified about the boy, the chance that the family has two boys gets closer to 50 percent.

If the question was specific enough to make it absolutely clear who that boy is, then the chances of two boys would be exactly 50 percent. Likewise, if we state that the boy is the older (or younger) son, we are also making it absolutely clear who he is, and the chance of two boys is also 50 percent.

⑯ THE TWYNNE TWINS

The most likely position is in first place. Each position has an equally likely probability of having one of the twins. But when a twin is in first position, it is always the first, whereas when farther back there is a chance it is the second twin.

Let's run the numbers. Imagine that there are only three students: the Twynne twins and an Other. There are three equally likely ways the students can line up, from left to right:

TTO

TOT

OTT

In ⅔ cases the first twin is at the front of the line, and in ⅓ cases the first twin is in second place. (The first twin can never be in last place.)

With two Others, the six equally likely ways the four students can line up are:

TTOO

TOTO

TOOT

OTTO

OOTT

OTOT

In ³⁄₆, or ½ cases the first twin is at the front of the line; in ²⁄₆, or ⅓ cases the first twin is in second place; and in ⅙ cases the first twin is in third place. If you were to do the same with more and more Others, you would always discover that the most likely position for the first Twynne is first place. Likewise, the second twin is most likely to be in last place.

(In general, in a class of n people, the chance of the first twin being in first place is $n - 1$ divided by the sum of the numbers from 1 to $n - 1$. So in a class of 30 students the chance of a twin being in the first position is ²⁹⁄₄₃₅.)

⑪⑰ A SHOT OF MMMR

The answer is (2, 5, 5, 6, 7) and (3, 4, 5, 5, 8).

How do we get there? We know the median is 5, so we can fill in our first value: (X, X, 5, X, X)

If the range is 5, the highest and lowest values will be one of the following pairs: (0, 5), (1, 6), (2, 7), (3, 8), (4, 9), or (5, 10).

We can eliminate some of these pairs by considering the mean. If the mean is 5, the sum of all the numbers divided by 5 must equal 5, in other words, the sum of all the numbers must be 25. If the range is (0, 5) or (1, 6), then the numbers will always sum to less than 25. Likewise, if the range is (4, 9) and (5, 10), the numbers will always sum to more than 25. So we have two possible sets:

(2, X, 5, X, 7) and (3, X, 5, X, 8)

In the former case the two unknowns must add up to 11. We can eliminate 4 and 7, since this would make 7 the mode. So, these unknowns must be 5 and 6.

In the latter case the two unknowns must add up to 9. We can eliminate 3 and 6 since that would make 3 the mode. So these unknowns must be 4 and 5.

⑪⑧ LIES AND STATISTICS

Imagine that everyone in the elementary school has a grade F, and that everyone in the high school has a grade C.

If the high school has one more pupil than the elementary school, then the median will be a C.

If everyone who is an F improves to a D, and everyone who is a C improves to a B, and two new pupils whose grade is D or lower join the elementary school, everyone's grades will have improved and the median will now be D.

⑪⑨ THE LONELINESS OF THE LONG-DISTANCE RUNNER

Using only the information presented in the question the best estimate is 251 runners.

By looking out of the window you have confirmed that there are at least 251 runners. If there are exactly 251, then the probability that you would have seen this runner is $\frac{1}{251}$. If there are 252 runners, then the probability you would have seen this runner is $\frac{1}{252}$, which is a tiny bit less likely. The estimate that makes your guess the most likely option is that there are 251 runners.

Of course, you may have extra information that can help you make a more informed estimate, such as knowledge of your local race that you picked up from the radio or friends, or an assumption that races tend to have a round number of contestants. But without any of this information, the best guess is 251.

Statisticians call this method of deduction "maximum likelihood estimation."

(120) THE FIGHT CLUB

Beast, Mouse, Beast.

It's a counterintuitive answer, since your best course of action is to fight the strongest opponent twice.

If you had to win a *single* bout, you would choose to go against Mouse. He is the weaker fighter, less good than you, and it's likely you would win.

Similarly, if you were fighting *three* bouts, and you wanted to win as many as possible, you would choose to fight Mouse—the weakest—twice and Beast only once.

However, the question asks about maximizing your chances of winning *two fights in a row*. In this case, it is in your best interests to choose the option where you fight the tougher opponent twice, because if you have to win two bouts in a row, you must win the middle one. Your best course of action, therefore, is to choose the easiest adversary for the middle bout, which gives you two chances to beat the harder opponent in the other bouts.

We can prove this more rigorously using the numbers given.

If the probability of beating Beast is $\frac{2}{5}$ and of beating Mouse is $\frac{9}{10}$, we can draw up the following table. We work out the probabilities of each possible combination of winning two in a row, and add them up.

Beast	Mouse	Beast	Probability
Win	Win	Win	$\frac{2}{5} \times \frac{9}{10} \times \frac{2}{5} = \frac{36}{250}$
Win	Win	Lose	$\frac{2}{5} \times \frac{9}{10} \times \frac{3}{5} = \frac{54}{250}$
Lose	Win	Win	$\frac{3}{5} \times \frac{9}{10} \times \frac{2}{5} = \frac{36}{250}$
		Total	$\frac{126}{250}$

Mouse	Beast	Mouse	Probability
Win	Win	Win	$\frac{9}{10} \times \frac{2}{5} \times \frac{9}{10} = \frac{81}{250}$
Win	Win	Lose	$\frac{9}{10} \times \frac{2}{5} \times \frac{1}{10} = \frac{9}{250}$
Lose	Win	Win	$\frac{1}{10} \times \frac{2}{5} \times \frac{9}{10} = \frac{9}{250}$
		Total	$\frac{99}{250}$

In the first combination the total odds are $^{126}/_{250}$, which is just over half, so you are more likely than not to win two bouts in a row. In the second combination the odds are under half, so you are more likely to fail to win two bouts in a row.

The moral of the story is that winning the most battles is not always the best way to win the war.

⑫ TYING THE GRASS AND TYING THE KNOT

It is more likely.

Once the top ends are joined in pairs, the blades can be arranged as so:

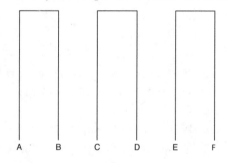

In order to make a loop, end A can be joined to any end except B, meaning that there is a choice of four out of the five options. Once A is joined to a permitted end, say C, then B can join to anywhere except D, since this will close a loop of four blades of grass. In other words, there are two out of three options for B. Let's say B is joined to E. This means that the final pair must be F and D. So the chances of creating a loop are $^4/_5 \times ^2/_3 = ^8/_{15}$, which is just over half.

⑫ THE THREE SLIPS OF PAPER

Let's say the three numbers are A, B, and C, where A > B > C.

The best strategy is to turn over two slips. If the second slip you turn over has a number that is bigger than the first, choose the second slip, but if the second slip has a lower number than the first, choose the remaining slip. This increases your chance of picking the slip with the highest number to 1 in 2.

We can see this by looking at the table below. If Slip 1 is the first slip you choose, and Slip 2 the second, the table lists the six equally likely arrangements of A, B, and C, together with what your choice will be. In three out of the six arrangements you will choose the highest number.

Slip 1	Slip 2	Slip 3	Choice
A	B	C	C
A	C	B	B
B	A	C	A
B	C	A	A
C	A	B	A
C	B	A	B

The solution for the problem when there are only two slips of paper is similar to the solution with three slips, only now we must randomly generate a number for an imaginary third slip. Here's how it works.

Let A and B be the numbers on the slips, with A > B.

We turn over our first slip. Now randomly generate a number, N.

Our strategy is now as follows: If N is bigger than the number on the first slip, choose the second slip, but if N is lower than the number on the first slip, choose the first slip.

There are three possibilities for N. It could be bigger than both A and B, in which case there is a 1 in 2 chance of choosing the correct slip, as shown below.

N > A > B

Slip 1	Slip 2	Choice
A	B	N > A, so B
B	A	N > B, so A

Or N is smaller than both A and B, in which case there is still a 1 in 2 chance of choosing the correct slip:

A> B > N

Slip 1	Slip 2	Choice
A	B	N < A, so A
B	A	N < B, so B

However, if N is *between* A and B, then the strategy gets us the right slip every time:

A > N > B

Slip 1	Slip 2	Choice
A	B	N < A, so A
B	A	N > B, so A

If there is a nonzero chance that our randomly generated number N is between A and B, our chances of picking the right number are better than 1 in 2. And there must be a nonzero chance, since whatever N is there will always be numbers on either side of it, and these could be A and B.

The solution to the problem when there are more than three slips of paper is to turn over the first 37 percent of the slips, and then choose the first slip with a number higher than any you have seen before. The percentage is the number $\frac{1}{e}$, where e is the exponential constant, which is 2.718 to three decimal places. Unfortunately, the proof is too detailed to include here.

⑫㉓ THE THREE PRISONERS

Neither of the prisoners reasoned correctly. The odds of A getting the pardon remain 1 in 3, but the odds of C getting the pardon increase to 2 in 3.

At the beginning of the problem, the governor picks one of the prisoners at random to be pardoned, which means that each prisoner has a 1 in 3 chance of being pardoned.

When the governor tells A that B will be executed, A's chances of being pardoned remain at 1 in 3, because whichever prisoner the governor decided to pardon, she will always be able to give A the name of another prisoner who will be executed. The fact that she said B in this case provides A with no useful information about his fate.

However, if the chances of A being pardoned remain at 1 in 3, the chances of A *not* being pardoned must be 2 in 3. In other words, the chance of B or C being pardoned is 2 in 3. However, we know that B will be executed. So the chance of C being pardoned must be 2 in 3.

(124) THE MONTY FALL PROBLEM

It's not in your advantage to switch. The chance of the car being behind door No. 1 and the chance of it being behind door No. 3 is 1 in 2 in both cases.

Let's look at this by drawing up a table of equally likely outcomes, and seeing what happens when you stick or switch. You have chosen door No. 1. There are six equally likely cases to consider, since there are two possible ways that Monty can fall for each possible location of the car.

Door concealing car	Door Monty opens	Stick	Switch
1	2	Win	Lose
1	3	Win	Lose
2	2	*	*
2	3	Lose	Win
3	2	Lose	Win
3	3	*	*

When accounting for all possible outcomes, we must include the times when Monty falls and opens the door concealing the car. I've marked these in the table with a *. We can ignore these rows in our calculations, however, since we know that the door opened to reveal a goat.

We can see that if Monty falls and opens a door concealing a goat, we will win by sticking in 2 cases out of 4, and win by switching in 2 cases out of 4. In other words, there is no advantage in switching.

The process by which Monty opens the door is crucial. In the original Monty

Hall problem, he knows where the car is, and his intention is to open a door that reveals a goat. This setup means that it's in your advantage to switch. In the Monty Fall problem, however, Monty arbitrarily opens a door that just happens to conceal a goat. To work out the correct probabilities, we need to take account of all the times he could have arbitrarily opened a door to reveal the car.

(125) RUSSIAN ROULETTE

In adjacent chambers, it's best to stick. In nonadjacent chambers, it's best to spin.

There are six chambers. Let's call them 1, 2, 3, 4, 5, and 6.

The adjacent case: The bullets are, say, in 1 and 2. When the cylinder is spun there is a ⅔ chance it will stop with a bullet-free chamber lined up with the barrel. If your captor spins again, the chances of getting a bullet-free chamber remain ⅔, or 66 percent. If you stick, however, and the cylinder turns to the next chamber, positions 3, 4, 5, and 6 become positions 4, 5, 6, and 1. Three of those four are bullet-free chambers, giving a chance of survival of 75 percent. You should stick.

The nonadjacent case: The bullets are, say, in 1 and 4. If you stick now, the bullet-free positions 2, 3, 5, and 6 become 3, 4, 6, and 1. Only two of those four are bullet-free chambers, giving you only a 50 percent chance of survival. Since spinning gives you a 66 percent chance of survival, you should spin.

BACK COVER

A list of the puzzles and a note on their sources

The list below contains the sources—books, magazines, websites, and friends—for the puzzles in this book. Some of the texts listed are not the original sources; it's often hard to find out exactly where a puzzle first emerged. If an original source is known, it can usually be found in David Singmaster's extensive and thorough *Sources in Recreational Mathematics*, available online, which I consulted on almost a daily basis. Every attempt has been made to contact copyright holders. All queries should be addressed to the publisher.

Tasty teasers
Number conundrums

(1) Ian Stewart, *Professor Stewart's Hoard of Mathematical Treasures*, Profile Books, 2009.

(2) Martin Gardner, *The Unexpected Hanging and Other Mathematical Diversions*, University of Chicago Press, 1969.

(3) Boris A. Kordemsky, *The Moscow Puzzles*, Dover Publications, 1992.

(4)–(6) Nobuyuki Yoshigahara, *Puzzles 101*, A. K. Peters/CRC Press, 2004.

The puzzle zoo
ANIMAL PROBLEMS

(1) The Three Rabbits. Traditional.

(2) Dead or Alive. *The Family Friend*, 1849.

(3) Good Neighbors. Des MacHale and Paul Sloane, *Hall of Fame Lateral Thinking Puzzles*, Sterling, 2011.

(4) A Fertile Family. Based on an idea from www.bio.miami.edu/hare/scary.html.

(5) A Bunch of Hops. Ron Knott, www.maths.surrey.ac.uk/hosted-sites/R.Knott/Fibonacci/fibpuzzles.html.

(6) Crossing the Desert. Adapted from Pierre Berloquin, *The Garden of the Sphinx*, Barnes & Noble, 1996.

(7) Save the Antelope. Adapted from Pierre Berloquin, *The Garden of the Sphinx*, Barnes & Noble, 1996.

(8) The Thirteen Camels. David Singmaster, *Sources in Recreational Mathematics*, South Bank University, 1991.

(9) Camel vs Horse. Traditional.

(10) The Zig-Zagging Fly. Traditional.

(11) The Ants on a Stick. Told to me by Rob Eastaway. www.robeastaway.com.

(12) The Snail on the Rubber Band. Martin Gardner, *Time Travel and Other Mathematical Bewilderments*, W. H. Freeman, 1988.

(13) Animals that Turn Heads. Kobon Fujimura, *The Tokyo Puzzles*, Frederick Muller, 1979. Nobuyuki Yoshigahara, *Puzzles 101*, A. K. Peters/CRC Press, 2004.

(14) Banishing Bugs from the Bed. Peter Winkler, *Mathematical Mind-Benders*, A. K. Peters/CRC Press, 2007.

(15) The Dumb Parrot. Yuri B. Chernyak and Robert M. Rose, *The Chicken from Minsk*, Basic Books, 1995.

(16) Chameleon Carousel. Question first posed in the International Tournament of the Towns, 1984.

(17) The Spider and the Fly. Henry Ernest Dudeney, *536 Curious Problems & Puzzles*, Barnes & Noble, 1995.

(18) The Meerkat in the Mirror. First told to me by Carlos Vinuesa.

(19) Catch the Cat. First told to me by Charlie Gilderdale.

(20) Man Spites Dog. Des MacHale and Paul Sloane, *Hall of Fame Lateral Thinking Puzzles*, Sterling, 2011.

(21) The Germ Jar. Naoki Inaba, "Numberplay" column, *The New York Times*.

(22) The Fox and the Duck. Martin Gardner, *Mathematical Carnival*, The Mathematical Association of America, 1989.

(23) The Logical Lions. Derrick Niederman, *Math Puzzles for the Clever Mind*, Puzzlewright Press, 2013.

(24) Two Pigs in a Box. Steven E. Landsburg, *Can You Outsmart an Economist?*, Mariner Books, 2018.

(25) Ten Rats and One Thousand Bottles. First heard on the YouTube channel *PBS Infinite Series*.

Tasty teasers
Grueling grids

(1), (2) Carlos D'Andrea, University of Barcelona.

(3) www.wesolveproblems.org.uk.

(4)–(6) Daniel Finkel, www.mathforlove.com.

I'm a mathematician, get me out of here
SURVIVAL PROBLEMS

㉖ Fire Island. Richard Wiseman's *Friday Puzzle*. richardwiseman.wordpress. com/2012/07/02/6488.

㉗ The Broken Steering Wheel. Adapted from Rob Eastaway and David Wells, *100 Maddening, Mindbending Puzzles*, Portico, 2018.

㉘ Walk the Plank. Adapted from Henry Dudeney, *536 Curious Problems & Puzzles*, Barnes & Noble, 1995.

㉙ The Three Boxes. Adapted from Raymond Smullyan, *What Is the Name of This Book?*, Dover Publications, 1978.

㉚ Safe Passage. Simon Singh, *The Code Book*, Fourth Estate, 1999.

㉛ Crack the Code. puzzling.stackexchange.com/questions/46871/crack-the-lock-code.

㉜ Guess the Password. "Technical Problems," from MIT's *The Tech*, April 17, 2005.

㉝ The Spinning Switches. Peter Winkler, *Mathematical Puzzles, A Connoisseur's Collection*, A. K. Peters/CRC Press, 2004.

㉞ Protect the Safe. Pierre Berloquin, *The Garden of the Sphinx*, Barnes & Noble, 1996.

㉟ The Secret Number. Steven E. Landsburg, *Can You Outsmart an Economist?*, Mariner Books, 2018.

㊱ Removing the Handcuffs. Martin Gardner, *Mathematics, Magic and Mystery*, Dover Publications, 1956.

㊲ The Reversible Pants. Martin Gardner, *Sixth Book of Mathematical Diversions from Scientific American*, University of Chicago Press, 1984.

㊳ Mega Area Maze. Naoki Inaba.

㊴ Arrow Maze. *Mathematical Olympiads 1999–2000: Problems and Solutions from Around the World*, Mathematical Association of America, 2002.

㊵ The Twenty-Four Guards. Adapted from Raymond Smullyan, *What Is the Name of This Book?*, Dover Publications, 1978.

㊶ The Two Envelopes. *Futility Closet*, 2009, www.futilitycloset.com/2009/08/05/royal-pain.

㊷ The Missing Number. Peter Winkler, *Mathematical Puzzles, A Connoisseur's Collection*, A. K. Peters/CRC Press, 2004.

㊸ The One Hundred Challenge. Rob Eastaway and David Wells, *100 Maddening, Mindbending Puzzles*, Portico, 2018.

㊹ The Fork in the Road. Martin Gardner, *My Best Mathematical and Logic Puzzles*, Dover Publications, 1994.

㊺ Bish and Bosh. The Fork in the Road. Martin Gardner, *My Best Mathematical and Logic Puzzles*, Dover Publications, 1994.

㊻ The Last Request. Raymond Smullyan, *The Riddle of Scheherazade*, A. A. Knopf, 1997.

(47) The Red and Blue Hats. "Mathematics in Education and Industry," *Maths Item of the Month*, August 2010. mei.org.uk/month_item_10#aug.

(48) The Majority Report. Peter Winkler, *Mathematical Puzzles, A Connoisseur's Collection*, A. K. Peters/CRC Press, 2004.

(49) The Room with the Lamp. Peter Winkler, *Mathematical Puzzles, A Connoisseur's Collection*, A. K. Peters/CRC Press, 2004.

(50) The One Hundred Drawers. Peter Winkler, *Mathematical Mind-Benders*, A. K. Peters/CRC Press, 2007.

Tasty teasers
Riotous riddles

(1) Traditional.

(2) Raymond Smullyan.

(3)–(10) Adapted from *Hall of Fame Lateral Thinking Puzzles*, by Sloane and MacHale, Sterling, 2011.

Cakes, cubes, and a cobbler's knife
GEOMETRY PROBLEMS

(51) The Box of Calissons. Guy David and Carlos Tomei, "The Problem of the Calissons," *The American Mathematical Monthly*, vol. 96, 1989.

(52) The Nibbled Cake. Source unknown.

(53) Cake for Five. Source unknown.

(54) Share the Doughnut. www.mathsisfun.com/puzzles/horace-and-the-doughnut.html.

(55) A Star Is Born. Edward B. Burger, *Making Up Your Own Mind*, Princeton University Press, 2018.

(56) Squaring the Rectangle. A version appears in the *Wakoku Chie-Kurabe*, 1727, and it has been repeated by many other authors since then.

(57) The Sedan Chair. *Mathematical Puzzles of Sam Loyd*, Dover, 1959.

(58) From Spade to Heart. *Mathematical Puzzles of Sam Loyd*, Dover, 1959.

(59) The Broken Vase. Pierre Berloquin, *The Garden of the Sphinx*, Barnes & Noble, 1996.

(60) Squaring the Square. Derrick Niederman, *Math Puzzles for the Clever Mind*, Puzzlewright Press, 2013.

(61) Mrs. Perkins's Quilt. Henry Ernest Dudeney, *Amusements in Mathematics*, 1917.

(62) The Sphinx and Other Reptiles. Author's own.

(63) Alain's Amazing Animals. en.tessellations-nicolas.com.

64 The Overlapping Squares. Pierre Berloquin, *The Garden of the Sphinx*, Barnes & Noble, 1996.

65 The Cut-Up Triangle. Nobuyuki Yoshigahara, *Puzzles 101*, A. K. Peters/CRC Press, 2004.

66 Catriona's Arbelos. Catriona Shearer. twitter.com/cshearer41.

67 Catriona's Cross. Catriona Shearer. twitter.com/cshearer41.

68 Cube Angle. Kobon Fujimura, *The Tokyo Puzzles*, Frederick Muller, 1979.

69 The Menger Slice. As told to me by George Hart. www.georgehart.com.

70 The Peculiar Peg. Martin Gardner, *The Second Scientific American Book of Mathematical Puzzles and Diversions*, University of Chicago Press, 1961.

71 The Two Pyramids. Peter Winkler, *Mathematical Puzzles, A Connoisseur's Collection*, A. K. Peters, 2004.

72 The Rod and the String. *Trends in International Mathematics and Science Study*, 1995.

73 What Color Is the Beard? Author's own.

74 Around the World in 18 Days. Jules Verne, *Around the World in Eighty Days*, 1873.

75 A Whiskey Problem. http://mathforum.org/wagon/2014/p1191.html.

Tasty teasers
Pencils and utensils

1 Fredrik Cattani, via email. www.filiokusfredrik.se.

2 Pierre Berloquin, *100 Games of Logic*, Barnes & Noble, 1977.

3 "Good Hands," *Futility Closet*, 2012. www.futilitycloset.com/2012/04/30/good-hands/.

4 Fredrik Cattani, via email, www.filiokusfredrik.se.

5 Source unknown.

A wry plod
WORD PROBLEMS

76 The Sacred Vowels. Adapted from Christian Bök, *Eunoia*, Canongate, 2009.

77 Winter Reigns. *Word Ways*, February, 1974.

78 Five Deft Sentences. 1, 3, 5. Author's own. 2. www.grammarly.com/blog/16-surprisingly-funny-palindromes. 4. Dmitri Borgmann, *Language on Vacation*, Charles Scribner's Sons, 1965.

79 The Consonant Gardener. Author's own and Martin Gardner, *Sixth Book of Mathematical Diversions from Scientific American*, University of Chicago Press, 1984.

80 Kangaroo Words. Chris Smith's *Maths Newsletter*, issue 476.

81 The Ten-Letter Words. Author's own.

(82) Ten Notable Numbers. Sources unknown.

(83) The Questions That Count Themselves. Lee Sallows. www.leesallows.com.

(84) The Sequence That Describes Itself. Eric Angelini.

(85) Sexy Lexy. Based on an idea from Eric Angelini.

(86) Letters in a Box. Martin Gardner, *Sixth Book of Mathematical Diversions from Scientific American*, University of Chicago Press, 1984; Peter Winkler, *Mathematical Mind-Benders*, A. K. Peters/CRC Press, 2007.

(87) Wonderful Words. Various sources, including Martin Gardner, *Mind-Boggling Word Puzzles*, Sterling, 2001; and Henry Ernest Dudeney, *The World's Best Word Puzzles*, 1925.

(88) Life Sentences. 1, 2. Widely known. 3. *More Mathematical Puzzles of Sam Loyd*, Dover Publications, 1960.

(89) In the Beginning (and the Middle and the End) Was the Word. Martin Gardner, *Mind-Boggling Word Puzzles*, Sterling, 2001.

(90) Looking at Letters. Peter Winkler, *Mathematical Mind-Benders*, A. K. Peters/CRC Press, 2007, and Martin Gardner, *Mind-Boggling Word Puzzles*, Sterling, 2001.

(91) A Matter of Reflection. Adapted from Rob Eastaway and David Wells, *100 Maddening, Mindbending Puzzles*, Portico, 2018.

(92) The Blank Column. Martin Gardner, *Wheels, Life and Other Mathematical Amusements*, W. H. Freeman, 1983.

(93) Welcome to the Fold. Scott Kim.

(94) My First Ambigram. Scott Kim.

(95) Boxed Proverbs. Scott Kim.

(96) Nmrcl Abbrvtns. Edwin F. Meyer and Joseph R. Luchsinger, *Book of Puzzles*, Gedanken Publishing, 2012.

(97) The Name of the Father. *200 Problems in Linguistics and Mathematics*, 1972, as translated in Tanya Khovanova's *Math Blog*, blog.tanyakhovanova.com.

(98) Telling the Time in Tallinn. Adapted from Babette Newsome, *North American Computational Linguistics Olympiad*, 2014, online practice problem, www.nacloweb.org.

(99) Counting in the Rainforest. Adapted from Dragomir Radev, *North American Computational Linguistics Olympiad*, 2012, online practice problem, www.nacloweb.org.

(100) Chemistry Lesson. www.lingling.ru, as translated in Tanya Khovanova's *Math Blog*, blog.tanyakhovanova.com.

Tasty teasers
Bongard bafflers

Sample: Bongard Problem 6 by Mikhail Bongard.

(1) Bongard Problem 40 by Mikhail Bongard.

(2) Bongard Problem 44 by Mikhail Bongard.

(3) Bongard Problem 29 by Mikhail Bongard.

(4) Bongard Problem 180 by Harry Foundalis.

Sleepless nights and sibling rivalries
PROBABILITY PROBLEMS

(101) Better Than Half a Chance. Edward B. Burger, *Making Up Your Own Mind*, Princeton University Press, 2018.

(102) Single White Pebble. Unknown source.

(103) The Joy of Socks. Raymond Smullyan, *What Is the Name of This Book?*, Dover Publications, 1978.

(104) Loose Change. *Half a Century of Pythagoras Magazine*, The Mathematical Association of America, 2015.

(105) The Sack of Potatoes. *Half a Century of Pythagoras Magazine*, The Mathematical Association of America, 2015.

(106) The Bags of Candies. Peter Winkler, *Mathematical Mind-Benders*, A. K. Peters/CRC Press, 2007.

(107) A Strategy for the Displacement of Improper Thoughts. Lewis Carroll, *Pillow Problems*, Dover Publications, 1958.

(108) Bertrand's Box Paradox. Joseph Bertrand, *Calcul des probabilités*, 1889.

(109) The Dice Man Diet. Steven E. Landsburg, *Can You Outsmart an Economist?*, Mariner Books, 2018.

(110) Die! Die! Die! Peter Winkler, *Mathematical Puzzles: A Connoisseur's Collection*, A. K. Peters/CRC Press, 2004.

(111) The Phony Flips. Author's own.

(112) Just Four Kids. *Futility Closet*, 2013, www.futilitycloset.com/2013/11/10/brood-war.

(113) The Big Family. Author's own.

(114) Problems with Siblings. Michael and Thomas Starbird.

(115) The Girl Born in an Even Year. Author's own.

(116) The Twynne Twins. Martin Gardner, *Wheels, Life and Other Mathematical Amusements*, W. H. Freeman, 1983.

(117) A Shot of MMMR. Adapted from *NRICH*, nrich.maths.org/11281.

(118) Lies and Statistics. Adapted from Steven E. Landsburg, *Can You Outsmart an Economist?*, Mariner Books, 2018.

(119) The Loneliness of the Long-Distance Runner. Steven E. Landsburg, *Can You Outsmart an Economist?*, Mariner Books, 2018.

(120) The Fight Club. Frederick Mosteller, *Fifty Challenging Problems in Probability*, Dover Publications, 1987.

(121) Tying the Grass and Tying the Knot. Martin Gardner, *Sixth Book of Mathematical Diversions from Scientific American*, University of Chicago Press, 1984.

(122) The Three Slips of Paper. Martin Gardner, *My Best Mathematical and Logic Puzzles*, Dover Publications, 1994.

(123) The Three Prisoners. Martin Gardner, *The Second Scientific American Book of Mathematical Puzzles and Diversions*, University of Chicago Press, 1961.

(124) The Monty Fall Problem. Jason Rosenhouse, *The Monty Hall Problem*, Oxford University Press, 2009.

(125) Russian Roulette. William Poundstone, *How Would You Move Mount Fuji?*, Little, Brown, 2003.

ACKNOWLEDGMENTS

I owe most thanks to the readers of my *Guardian* puzzle column.* I've been setting a puzzle every two weeks since May 2015, and thanks to readers' enthusiasm, suggestions and (helpful!) pedantry, I am constantly learning new, interesting problems, as well as always being kept on my toes.

Friends in the mathematical and puzzle community who have helped me with material include Carlos D'Andrea, Rob Eastaway, Tanya Khovanova, Max Maven, Adrián Paenza, Simon Pampena, Bernardo Recamán, Adam Rubin, Steve Selvin, Chris Smith, and Carlos Vinuesa. I'm grateful to Eric Angelini, Dan Finkel, Scott Kim, Lee Sallows, and Tom and Michael Starbird for letting me use problems they created, and to the authors Doug Nufer and Mike Keith for letting me reprint their constrained prose. It is a privilege to be able to use Scott's puzzle on the back cover.† Thanks to tessellation artist Alain Nicolas for giving me permission to use his artwork.‡ And thanks to Catriona Shearer for her geometrical puzzles. You can find many more on her Twitter, @Cshearer41.

* www.theguardian.com/science/series/alex-bellos-monday-puzzle

† For readers interested in ambigrams, his website www.scottkim.com has lots of further activities and examples.

‡ Readers wanting to see more of his amazing images should go to en.tessellations-nicolas.com.

In writing about Georges Perec, I am treading on the toes of my father David, who is Perec's biographer and translator and an expert on Oulipo. For my research about writing constraints, he recommended the essay "Exercises in Wile" by his Princeton University colleague Joshua Katz, from which I took the anagram that is the title of my chapter on wordplay.

My brilliant editors Fred Baty and Laura Hassan at Guardian Faber have coped fantastically well with having their brains turned inside out by these problems. It has also been a pleasure working with their colleagues Kate Ward in pre-press, Pete Adlington in design, Jack Murphy in production, and Josh Smith in publicity.

Ian Fitzgerald has managed the complicated process of putting the book together masterfully. Andri Johannsson drew perfect mathematical drawings, and Simon Landrein created marvelous illustrations for the beginning of each chapter. The text is immeasurably better thanks to Ben Sumner's razor-sharp copy edit, and looks gorgeous thanks to Richard Carr's design and typesetting.

As always, I have been superbly agented by Rebecca Carter and her colleagues Kirsty Gordon and Ellis Hazelgrove at Janklow & Nesbit. I'm grateful to the ever-perceptive and insightful Colin Beveridge and Moses Klein for giving the book a mathematical once-over.

Lastly, I'd like to thank Natalie for her support, patience, and cakes.

ABOUT THE AUTHOR

ALEX BELLOS is brilliant on all things mathematical. He has a degree in mathematics and philosophy from Oxford University. His bestselling books, *Here's Looking at Euclid* and *The Grapes of Math,* have been translated into more than twenty languages and were both short-listed for the Royal Society Science Books Prize. He is also the author of *Can You Solve My Problems?* and *Puzzle Ninja* and the coauthor, with Edmund Harriss, of the mathematical coloring books *Patterns of the Universe* and *Visions of the Universe*. He has launched an elliptical pool table, LOOP. He writes a puzzle blog for *The Guardian*, and in 2016, he won the Association of British Science Writers award for best science blog.

AlexBellos.com | @alexbellos